Discovering Sociology

An
Introduction
Using
ExplorIt®

Steven E. Barkan

MicroCase®
CORPORATION

14110 NE 21st Street
Bellevue, WA 98007

DISCOVERING SOCIOLOGY: An Introduction Using ExplorIt is published by MicroCase Corporation.

Editor	David J. Smetters
Editorial Assistant	Julie Aguilar
Production Manager	Jodi B. Gleason
Software, Lead Developer	David H. Simmons
Software, Programmer	Dana Schlatter
Data Archivists	Meredith Reitman
	Chris Bader
Copy Editors	Margaret Moore
	Judith Abrahms
Interior / Cover Design	Michael Brugman Design
Cover Photo	Todd Davidson/The Image Bank

CONTENTS

ABOUT THE AUTHOR

Steven E. Barkan is Professor of Sociology at the University of Maine, where he has taught Introduction to Sociology and many other courses since 1979. His teaching and research interests include criminology, research methods, sociology of law, and social movements. Among his professional activities, he has served as a member of the Advisory Board of the American Sociological Association's Honors Program, as the chair of the Law and Society Division of the Society for the Study of Social Problems, and as an advisory editor of that Society's journal, *Social Problems*. His previous books include: *Criminology: A Sociological Understanding* and *Protesters on Trial: Criminal Justice in the Southern Civil Rights and Vietnam Antiwar Movements*. He has also written numerous journal articles dealing with topics such as death penalty attitudes, political trials, feminist activism, and race and political participation. These articles have appeared in the *American Sociological Review*, *Journal of Research in Crime and Delinquency*, *Social Forces*, *Social Problems*, *Sociological Forum*, *Sociological Inquiry*, and *Race and Society*. Professor Barkan welcomes comments from students and faculty about this workbook. You may reach him by e-mail at BARKAN@MAINE.MAINE.EDU or by regular mail at: Department of Sociology, 5728 Fernald Hall, University of Maine, Orono, ME 04469-5728.

PREFACE

Sociological knowledge is generated by research of many types. Among the most common types is quantitative research, the analysis of numerical data. This workbook comes with three data sets, each containing real data that sociologists use in their own research. One of the data sets comes from a national survey of the U.S. population; another derives from the U.S. census and other measures of the 50 states; and a third includes international data on most of the world's nations and population.

With these data sets and the aid of MicroCase's ExplorIt software, this workbook will help you discover sociology by *doing* sociology. In every exercise, you will first see examples of data analysis on the standard topics central to the sociological discipline, and then you will have the chance to do your own analysis of data from the three data files. This active engagement with the sociological enterprise will help you learn about society in ways unimaginable little more than a decade ago.

My interest in writing this workbook stems from my first year in college in 1969, when I began learning sociology by analyzing numerical data contained on keypunch cards that I fed through a card sorter. However antiquated the terms "keypunch cards" and "card sorter" must sound to today's students, this active engagement with the process and logic of data analysis triggered my fascination with sociology. When MicroCase came along in the late 1980s, I took to it eagerly and began using it with my own students. The software has advanced considerably since that time, and what you now have in this workbook represents a unique combination of simplicity and sophistication in social science data analysis software. You will learn how to use the software in seconds, and you will explore hundreds of social relationships as you discover sociology.

I am delighted to acknowledge the efforts of several people and organizations without whom this workbook would not have been possible. Norman Miller, my first sociology professor, introduced me to sociology and the fascination of empirical research. Writing this workbook is one way of repaying the considerable debt I owe him. The MicroCase staff, and in particular Julie Aguilar, Jodi Gleason, and David Smetters, responded quickly to my many questions and in this and other respects provided much help and encouragement. They have a wonderful vision for introducing today's students to social science knowledge and data analysis, and I hope this workbook will help them fulfill that vision.

I also would like to thank the sources of the data files accompanying this workbook. MicroCase's data archive provided all three of these files, and their many hours involved in compiling these files saved me even more hours of labor. Special thanks go to Tom W. Smith of the National Opinion Research Center for his continued direction and administration of the General Social Survey. Much of the international data in the workbook is based on the World Values Survey, for which I thank Ronald Inglehart at the Institute for Social Research, University of Michigan. Professor Inglehart's book, *Modernization and Postmodernization: Cultural, Economic and Political Change in 43 Societies* (Princeton University

Press, 1997) inspired several of the examples in the workbook. Thanks also go to the Roper Center for its many years of distribution of the General Social Survey and to the Inter-university Consortium for Political and Social Research (ICPSR). The data archives maintained by ICPSR, the Roper Center, and MicroCase Corporation provide an invaluable service to social science researchers and their students.

My next set of acknowledgments goes to the several instructors who reviewed the manuscript. Their comments were critical but fair and greatly improved the final product. Needless to say, any remaining errors are my responsibility. The reviewers are: William Clute, University of Nebraska–Omaha; Charles Faupel, Auburn University; James D. Jones, Mississippi State University; Josephine Ruggiero, Providence College; Lynne Schlesinger, SUNY at Plattsburgh; James Sherohman, St. Cloud State University; Bill Tregea, Adrian College; and James Williams, University of Wisconsin–Eau Claire. Special thanks go to Kevin Demmitt at Clayton State College for class-testing a preliminary version of the workbook and software.

My final acknowledgment goes to my family, Barb, Dave, and Joey. I dedicate this book to them for their love, patience, and understanding as I juggled the writing of this workbook with my other professional and family responsibilities. I hope they will agree that the effort was worth it.

GETTING STARTED

INTRODUCTION

Welcome to ExplorIt! With the easy-to-use software accompanying this workbook, you will have the opportunity to learn about sociology by exploring dozens of sociological issues with data from the United States and around the world.

Each exercise in this workbook deals with a standard topic in typical sociology courses. The preliminary section of each exercise uses data from the workbook's software to illustrate key issues and social relationships related to that topic. You can easily create all the graphics in this part of the exercise by following the ExplorIt Guides you'll be seeing. Doing so will take just a few clicks of your computer mouse and will help you become familiar with ExplorIt. The ExplorIt Guides are described in more detail below.

Each exercise also has a worksheet section where you'll do your own data analysis. This section usually contains about a dozen questions that will either follow up on examples from the preliminary section or have you explore some new issues. You'll use the workbook's software to answer these questions.

SYSTEM REQUIREMENTS

Two versions of Student ExplorIt have been provided with this book: a Windows 95 version and a DOS version. The Windows 95 version can be used only on computers running Windows 95 (or higher). The DOS version of the software will run on almost any DOS- or Windows-compatible computer, including those using Windows 3.1 and Windows 95. Here is more detailed information about the minimum computer requirements for each version of the student software.

> **Student ExplorIt for Windows 95**—This version will run on almost any computer using Windows 95 (or higher).[1] For installation purposes, a CD-ROM drive and a 3.5" floppy drive are also required.

> **Student ExplorIt for DOS** (or Windows 3.1)—This version requires an IBM or compatible computer with 286 or better processor, 640K RAM, DOS 3.1 or higher (or Windows 3.1 or higher), VGA-level graphics, and a mouse. A CD-ROM drive is **not** required.

Before installing any software on a hard drive, check with your instructor to see if a network version of Student ExplorIt has already been installed. If a network version has already been installed, skip to the section "Starting Student ExplorIt."

If the operating system on your computer is DOS or Windows 3.1, then you must use *Student ExplorIt for DOS*. If this is your case, skip to the section "Starting Student ExplorIt." Even if your computer has

---

[1] *Student ExplorIt for Windows 95* requires 8 megabytes of RAM, 10 megabytes of free hard disk space (network installations require about 1 megabyte of temporary storage on hard drives of user terminals), VGA-level graphics, and a mouse.

Windows 95 (or higher), there are some situations in which you should still use *Student ExplorIt for DOS*:

- Your teacher has instructed you to use the DOS version (yes, the DOS version can be used on Windows 95 computers).

- You are not allowed to install software on the hard drive of the computer (such as in a lab setting).

- Your computer does not have a CD-ROM drive.

- Your computer does not have a CD-ROM drive and a floppy drive that can be used at the same time (as with some notebook computers).

If any of these conditions apply, you should skip to the section "Starting Student ExplorIt."

NETWORK VERSIONS OF STUDENT EXPLORIT

Network versions are available for both the *Student ExplorIt for Windows 95* and *Student ExplorIt for DOS*. These special versions of the software are available at no charge to instructors who adopt this book for their courses (instructors should contact MicroCase Corporation for additional information). It's worth noting that *Student ExplorIt for DOS* can be run directly from the diskette on virtually any computer network—regardless of whether a network version of Student ExplorIt has been installed.

INSTALLING STUDENT EXPLORIT FOR WINDOWS 95

If you will be using *Student ExplorIt for DOS* (see above discussion), you do not need to read this section. Skip to "Starting Student ExplorIt."

To install *Student ExplorIt for Windows 95*, you will need the diskette and CD-ROM that are packaged inside the back cover of this book. Then follow these steps:

1. Start your computer and wait until the Windows 95 desktop is showing on your computer.

2. Insert the diskette into the A drive (or B drive) of your computer.

3. Insert the CD-ROM disc into the CD-ROM drive.

4. Click [Start] from the Windows 95 desktop, click [Run], type **D:\SETUP** (if your CD-ROM drive is not the D drive, replace the letter D with the proper drive letter), and click [OK].

5. During the installation, you will be presented with several screens (described below). In some cases you will be required to make a selection or entry and then click [Next] to continue.

The first screen that appears is the **Welcome** screen. This provides some introductory information and suggests that you shut down any other programs that may be running. Click [Next] to continue.

You are next presented with a **Software License Agreement**. Read this screen and click [Yes] if you accept the terms of the software license.

If this is the first time you are installing Student ExplorIt, an **Install License** screen appears. (If this software has been previously installed or used, it already contains the licensing information. A screen simply confirming your name will appear instead.[2]) Here you are asked to type in your name. It is important to type your name correctly, since it cannot be changed after this point. Your name will appear on all printouts, so make sure you spell it completely and correctly! Then click [Next] to continue.

The next screen has you **Choose the Destination** for the program files. You are strongly advised to use the destination directory that is shown on the screen. Click [Next] to continue.

The **Install Checkbox** screen requires you to make a choice as to whether or not to copy the data files (currently located on the diskette) to your hard drive. Carefully read the choices on the screen before making your selection.

When the **Setup Complete** window appears, click [Finish]. You will find it easier to start Student ExplorIt if you place a "shortcut" icon on your Windows desktop. A folder named "MicroCase" should now be showing on the horizontal task bar at the bottom of your Windows desktop. Click on this button and a window will appear with a "shortcut" icon for Student ExplorIt.[3] Place your mouse pointer over this icon, then press down *and* hold the left mouse button as you drag the icon outside the window to an open space on your Windows desktop. Once the Student ExplorIt icon has been moved to your desktop, you can close the window that previously contained the icon by clicking the little "x" button that appears in the top right corner of the window. From this point on, you will be able to double-click the Student ExplorIt shortcut icon to start the software.

STARTING STUDENT EXPLORIT

The first section below describes how to start *Student ExplorIt for Windows 95*, while the second section describes how to start *Student ExplorIt for DOS*. Read the section that is appropriate for you.

Starting Student ExplorIt for Windows 95

Student ExplorIt for Windows 95 must be installed on a hard drive (or a computer network) before you can start it. If the program has not been installed, review the software installation section above.

If the data files were not copied to the hard drive of the computer during the installation of *Student ExplorIt for Windows 95*, it will be necessary for you to insert your 3.5" data file diskette into the A or B

---

[2] If an installation of Student ExplorIt is already on your hard drive, you will get a warning message indicating that a copy of the program is already present on your computer. If your intention is to *replace* the previously installed version of Student ExplorIt, use the default directory offered by the installation program. If you want to create a completely separate installation of Student ExplorIt, select a new directory using the "Browse" button.

[3] If you have another installation of Student ExplorIt on your hard drive (including a version distributed with a different MicroCase textbook), make sure to rename the label for its "shortcut" icon on your Windows desktop before following the instructions in the next sentence (you can name it anything except "Student ExplorIt"). To rename a shortcut, click once on the shortcut label, then click it again. This causes the text for the shortcut to be highlighted, after which it can be modified.

drive of your computer. (If you are starting Student ExplorIt from a network, you *must* insert your diskette before continuing.) Don't worry, you will be prompted to insert your diskette if you forget.

If the software was installed properly, there should be a "shortcut" icon on your Windows desktop that looks something like this:

To start *Student ExplorIt for Windows 95*, position your mouse pointer over the shortcut icon and double-click (that is, click it twice in rapid succession). If you did not move the shortcut icon onto your desktop during the install process, you can alternatively follow these directions to start the software.

Click [Start] from the Windows 95 desktop.

Click [Programs].

Click MicroCase.

Click Student ExplorIt.

After a few seconds, Student ExplorIt should appear on your screen. Skip down to the "Main Menu of Student ExplorIt" section below to continue your introduction to the software.

Starting Student ExplorIt for DOS

This section explains how to start *Student ExplorIt for DOS*. You can run *Student ExplorIt for DOS* directly from the diskette on almost any DOS or Windows computer (including computers using Windows 3.1 or Windows 95).

The instructions for starting *Student ExplorIt for DOS* differ depending on the operating system you are using. In all cases, you will first need to place the 3.5" diskette in the A or B drive. Go ahead and do that now. Then follow the appropriate instructions below to start *Student ExplorIt for DOS* on your computer.

<u>**MS-DOS:**</u>

Type **A:EXPLORIT** (or **B:EXPLORIT**) and press <Enter>.

<u>**Windows 3.1:**</u>

From the Program Manager, click [File].

Click [Run].

Type **A:EXPLORIT** (or **B:EXPLORIT**) and click [OK].

<u>**Windows 95:**</u>

Click [Start].

Click [Run].

Type **A:EXPLORIT** (or **B:EXPLORIT**) and click [OK].

The first time you start *Student ExplorIt for DOS,* you will be asked to enter your name. It is important to type your name correctly, since it will appear on all printouts. Type your name and click [OK] or press <Enter>. If your name is correct, simply click [OK] or press <Enter> in response to the next prompt. (If you wish to correct a mistake, click [Cancel] to make a correction.) To continue to the main menu of the program, press the <Enter> key or click the left mouse button.

Note: If you are using Windows 3.0 or 3.1 and the mouse fails to appear or it does not work properly, refer to Appendix A.

MAIN MENU OF STUDENT EXPLORIT

Student ExplorIt is extremely easy to use. All you do is point and click your way through the program. That is, use your mouse arrow to point at the selection you want, then click the left button on the mouse. The main menu is the starting point for everything you will do in Student ExplorIt. Let's take a look at how it works.

Student ExplorIt for Windows 95—Not all options on the menu are always available. In the Windows 95 version of Student ExplorIt, you will know which options are available at any given time by looking at the colors of the options. For example, when you first start the software, only the OPEN FILE option is immediately available. As you can see, the colors for this option are brighter than those for the other tasks shown on the screen. Also, when you move your mouse pointer over this option, it is highlighted.

Student ExplorIt for DOS—When you are at the main menu of the DOS version of Student ExplorIt, only those tasks that have a bright yellow background are available. As you can see, no tasks are available until you select a data file with which to work.

EXPLORIT GUIDES

Throughout this workbook, there are "ExplorIt Guides" that provide you with the basic information needed to carry out each task. Here is an example:

➤ *Data File:* **STATES**
➤ *Task:* **Mapping**
➤ *Variable 1:* **97) MURDER**
➤ *View:* **Map**

Each line of the ExplorIt Guide is actually an instruction. Let's follow the simple steps to carry out this task.

Step 1: Select a Data File

Before you can do almost anything in Student ExplorIt, you need to open a data file.

Student ExplorIt for Windows 95—To open a data file, click the OPEN FILE task. A list of data files will appear in a window (e.g., GSS, NATIONS, STATES, etc.). If you click on a file name *once*, a description of the highlighted file is shown in the window next to this list. In the ExplorIt Guide shown above, the ➤ symbol to the left of the Data File step indicates that you should open the STATES data file. To do so, click STATES and then click the [Open] button (or just double-click STATES). The next window that appears (labeled File Settings) provides additional information about the data file, including a file description, the number of cases in the file, and the number of variables, among other things. To continue, click the [OK] button. You are now returned to the main menu of Student ExplorIt. (You won't need to repeat this step until you want to open a different data file.) Notice that you can always see which data file is currently open by looking at the file name shown on the top line of the screen.

Student ExplorIt for DOS—In the DOS version of Student ExplorIt, the available data files are listed in the window at the left of the screen, and the description of the highlighted file is shown in the window beneath this list. To see the description of a file, click it once. To select a file, double-click its name. The "x" in the box next to the name of the file indicates which file is open. In this example, you should open the STATES data file. (You won't need to repeat this step until you want to use a different data file.)

Step 2: Select a Task

Once you have selected a data file, the next step is to select a program task. Six analysis tasks are offered in this version of Student ExplorIt. Not all tasks are available for each data file, because some tasks are appropriate only for certain kinds of data. Mapping, for example, is a task that applies only to ecological data, and thus cannot be used with survey data files.

In the ExplorIt Guide we're following, the ➤ symbol on the second line indicates that the MAPPING task should be selected, so click the MAPPING option with your left mouse button.

Step 3: Select a Variable

After a task is selected, you will be shown a list of the variables in the open data file. Notice that the first variable is highlighted and a description of that variable is shown in the Variable Description window at the lower right. You can move this highlight through the list of variables by using the up and down cursor keys (as well as the <Page Up> and <Page Down> keys). You can also click once on a variable name to move the highlight and update the variable description. Go ahead—move the highlight to a few other variables and read their descriptions.

If the variable you want to select is not showing in the variable window, click on the scroll bars located on the right side of the variable list window to move through the list. See the following figure:

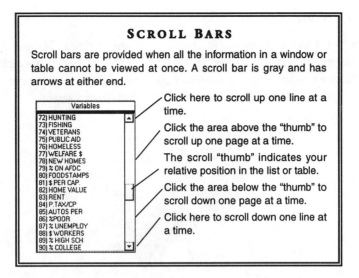

SCROLL BARS

Scroll bars are provided when all the information in a window or table cannot be viewed at once. A scroll bar is gray and has arrows at either end.

Click here to scroll up one line at a time.

Click the area above the "thumb" to scroll up one page at a time.

The scroll "thumb" indicates your relative position in the list or table.

Click the area below the "thumb" to scroll down one page at a time.

Click here to scroll down one line at a time.

By the way, you will find an appendix section at the back of this workbook (Appendix B) that contains a list of the variable names for key data files provided in this package.

Each task requires you to select one or more variables, and the ExplorIt Guides indicate which variables should be selected. The ExplorIt Guide example here indicates that you should select 97) MURDER as Variable 1. On the screen, there is a box labeled Variable 1. Inside this box, there is a vertical cursor that indicates that this box is currently an active option. When you select a variable, it will be placed in this box. Before selecting a variable, be sure that the cursor is in the appropriate box. If it is not, place the cursor inside the appropriate box by clicking the box with your mouse. This is important because in some tasks the ExplorIt Guide will require more than one variable to be selected, and you want to be sure that you put each selected variable in the right place.

To select a variable, use any one of the methods shown below. (Note: If the name of a previously selected variable is in the box, use the <Delete> or <Backspace> key to remove it—or click the [Clear All] button.)

- Type in the **number** of the variable and press <Enter>.

- Type in the **name** of the variable and press <Enter>. Or, you can type just enough of the name to distinguish it from other variables in the data—MUR would be sufficient for this example.

- Double-click the desired variable in the variable list window. This selection will then appear in the variable selection box. (If the name of a previously selected variable is in the box, the newly selected variable will replace it.)

- In *Student ExplorIt for Windows 95*, you have a fourth way to select a variable. First highlight the desired variable in the variable list, then click the arrow that appears to the left of the variable selection box. The variable you selected will now appear in the box. (If the name of a previously selected variable is in the box, the newly selected variable will replace it.)

Once you have selected your variable (or variables), click the [OK] button to continue to the final results screen.

Step 4: Select a View

The next screen that appears shows the final results of your analysis. In most cases, the screen that first appears matches the "view" indicated in the ExplorIt Guide. In this example, you are instructed to look at the Map view—that's what is currently showing on the screen. In some instances, however, you may need to make an additional selection to produce the desired screen.

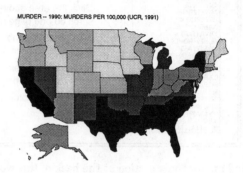

MURDER -- 1990: MURDERS PER 100,000 (UCR, 1991)

(OPTIONAL) Step 5: Select an Additional Display

Some ExplorIt Guides will indicate that an additional "Display" should be selected. In that case, simply click on the option indicated for that additional display. For example, this ExplorIt Guide may have included an additional line that required you to select the [Legend] display.

Step 6: Continuing to the Next ExplorIt Guide

Some instructions in the ExplorIt Guide may be the same for at least two examples in a row. For instance, after you display the map for murder in the example above, the following ExplorIt Guide may be given:

> Data File: **STATES**
> Task: **Mapping**
> ➤ Variable 1: **95) V.CRIME**
> ➤ View: **Map**

Notice that the first two lines in the ExplorIt Guide do not have the ➤ symbol located in front of the items. That's because you already have the data file STATES open and you have already selected the MAPPING task. With the results of your first analysis showing on the screen, there is no need to return to the main menu to complete this next analysis. Instead, all you need to do is select V.CRIME as your new variable. If you are using *Student ExplorIt for Windows 95*, click the [⤴] button located in the top left corner of your screen (if you are using *Student ExplorIt for DOS*, click the [Exit] button once). The variable selection screen for the MAPPING task appears again. Replace the variable with 95) V.CRIME and click [OK].

To repeat: You need only do those items in the ExplorIt Guide that have the ➤ symbol in front of them. If you start from the top of the ExplorIt Guide, you're simply wasting your time.

Getting Started

If the ExplorIt Guide instructs you to select an entirely new task or data file, you will need to return to the main menu. To return to the main menu using *Student ExplorIt for Windows 95*, simply click the [Menu] button located at the top left corner of the screen. (To return to the main menu using *Student ExplorIt for DOS*, click the [Exit] button until the main menu appears. At this point, select the new data file and/or task that is indicated in the ExplorIt Guide.)

That's all there is to the basic operation of Student ExplorIt. Just follow the instructions given in the ExplorIt Guide and point and click your way through the program.

EXITING FROM STUDENT EXPLORIT

If you are continuing to the next section of this workbook, it is *not* necessary to exit from Student ExplorIt quite yet. But when you are finished using the program, it is very important that you properly exit the software—do not just walk away from the computer or remove your diskette. To exit Student ExplorIt, return to the main menu and select the [Exit] button that appears on the screen.

Important: If you inserted your diskette before starting Student ExplorIt, remember to remove it before leaving the computer.

THE SOCIOLOGICAL PERSPECTIVE

Tasks: Mapping
Data Files: STATES

We are all individuals, but we are also social beings influenced by our social environments. We grow up in a society and within a family in that society. We grow up as girls or boys and, sooner than our parents would like, become women or men. We are members of a racial or ethnic group or, depending on our parents' ancestry, more than one such group. We have low incomes, medium incomes, or high incomes. Many of us belong to a religious faith. We also grow up in different parts of the country and in urban, suburban, or rural areas in the country. All these aspects of our social environment make up our social background, and they all influence the way we turn out: how we think, how we behave, and our chances for success or failure in life.

This is the fundamental truism of what is often called the sociological perspective: our social backgrounds influence our attitudes, behavior, and life chances. While no two people, even identical twins, are the same, neither are they completely different. If they've both grown up in the United States, they are automatically more similar to each other than if one had grown up in the U.S. and the other in Nigeria or Japan. If they are the same sex—both female or both male—they have more things in common than if one were female and the other male. All this means that if we know enough about an individual's background, we can predict her or his attitudes, behavior, and eventual outcomes in life with surprising accuracy. We won't always be right, but we will be right more often than we're wrong.

This workbook's major goal is to illustrate the sociological perspective with real-life data drawn from the United States and from around the world. You will learn sociology by doing sociology. You'll see again and again how social backgrounds influence behavior, attitudes, and life chances. You'll also see some surprising examples of expected influences not happening. Although much of our exploration will take place in the United States, we'll also be looking at a wide range of other nations, which differ in many ways that have important consequences for the attitudes, behavior, and life chances of their populations. Such a global perspective is increasingly important in today's world, not least because it helps us understand our own society better.

Welcome, then, to our use of Student ExplorIt to discover sociology. The computer program itself is easy and even fun to use, and the hundreds of variables contained in the data sets accompanying this workbook provide you with a vast quantity of intriguing data to aid in your discovery.

This first chapter introduces you to the sociological perspective by using maps of the United States to illustrate regional differences in behavior and possible explanations for these differences. Later chapters will follow a different format. In those, we will first examine the chapter's subject matter with international data, then continue with data on the states of the United States, and end with data from a large national survey of U.S. residents conducted in 1996. These three data sets will complement each other as you discover sociology.

SUICIDE AND THE SOCIOLOGICAL PERSPECTIVE

Suicide is one of the most tragic acts we can think of and also one of the hardest to explain. When we hear that someone committed suicide, a natural question to ask is, "Why did she kill herself?" (or "Why did he kill himself?"). When we try to answer this question, we usually speculate that the suicide victim had been depressed over some unfortunate circumstance in his or her life. A marriage or relationship was ending, a job or school work was going poorly, the person's health was failing, and the like. All these answers focus on the immediate, individual circumstances of the person's life, as well they should. What, after all, is a more individualistic act than suicide? What is more individualistic than deciding to take your own life?

A sociologist, however, approaches suicide as a social act, not just as an individual one. A sociologist asks, "Why is it that some *kinds* of individuals are more likely than other kinds to commit suicide?" To put it another way, "Why do some groups of individuals have higher *rates* of suicide than other groups?" The idea is that an individual from a group—or, more precisely, from a social background—with a higher rate of suicide is more likely to commit suicide than one from a group or social background with a low rate of suicide.

This central insight into the sociological perspective comes from the French sociologist Emile Durkheim, one of the founders of sociology. Writing a century ago, Durkheim said that explanations focusing on individual unhappiness are insufficient to explain differences in group suicide rates. Instead, he said, there must be something about the group itself that explains its high or low rate. Suicide, then, does not just arise from forces inside the individual; it also arises from forces external to the individual. These forces are the properties of the group to which the individual belongs or the social background from which the individual comes.

Durkheim's focus on the influence of social forces on individual behavior and attitudes greatly influenced the development of the sociological perspective. Let's look at the United States today to see if some of the suicide patterns that Durkheim would have predicted still hold true.

We'll begin by using your STATES data set. Like the other data sets that come with this workbook, the STATES data set contains many variables. In sociological jargon, a *variable* is anything that varies. To be more precise, it is any characteristic, feature, or dimension that takes on different values among the things (or *units of analysis*) being studied. In the STATES data set, the unit of analysis is the state. We have 50 states in this data set, or 50 cases. States differ in many ways, and the STATES data set contains information on many of the ways, or variables, in which states differ. These variables include the size of each state's population, the percent of each state that is poor, the percent of each state's population that is under the age of 5, and the circulation of *Field & Stream* magazine per 100,000 population. If we study individuals, the units of analysis in the GSS data set, we often study variables such as gender (either female or male), income (high, low, or medium), and political views (liberal, moderate, or conservative).

As the examples of state variables indicate, many of the variables in the STATES data set are either percents or rates. Both such figures take into account the fact that some states have more people than others. For example, to indicate how many poor people there are in California, a state of more than 30 million, and to compare that with the number of poor people in Kansas, a state of less than 3 million, tells us little about which state is poorer, because the states' population sizes are so different. Sociologists and other social scientists thus use percents or rates to compare geographic units such as states.

To calculate a *percent*, in case you have forgotten your early math lessons, we divide the number of people who are poor (to stick with our poverty example) by the number of people living in a state and multiply the result by 100. Thus, using some unrealistically small numbers to keep the math easier, if we have 25 poor people living in a state containing 200 people, we divide 25 by 200 to get .125 and then multiply this result by 100 to get 12.5 percent. This is equivalent to 12.5 poor people for every 100 poor people in the state, and we say that 12.5 percent of the state's population is poor.

A rate gives the same kind of information as a percent but commonly uses 1000 or 100,000 (if we're still looking at states) instead of 100. We use these larger multipliers when calculating a percent would yield percents less than 1. Thus, if the number of murders is 250 in a state of 2,300,000, dividing 250 by 2,300,000 gives us .0001086. If we only multiplied this figure by 100, we would say that .01086 percent of the state's population is murdered every year. Because murder is a rare event and thus yields percents much less than 1, we instead multiply .0001086 by 100,000 to give us a murder rate of 10.86 per 100,000 population. (Note that a poverty rate of 12.5 per 100 population, or 12.5 percent, is the same as a rate of 12,500 per 100,000 population.)

Back to business. One of the variables in the STATES data set is the number of suicides per 100,000 population for each state. Because this variable uses a rate and not just the number of suicides, we can rank the states according to their rates. The higher the suicide rate in a state, the more common suicide is in that state, and the more likely it is that an individual in that state will commit suicide.

I'll now show the ExplorIt Guide that lists the steps to obtain the map for each state's suicide rate. If you don't understand the Guide, reread the *Getting Started* section at the front of this book or the relevant section in Appendix A at the back of the book. When you see ExplorIt Guides in this and all other exercises in the book, you're strongly advised to actually do the steps at a computer. This procedure will help you learn how to use Student ExplorIt and become more familiar with the program. It will also help prepare you to complete the worksheets at the back of each exercise.

To repeat, our first ExplorIt Guide obtains a map depicting each state's suicide rate. Here we go!

➤ *Data File:* **STATES**
  ➤ *Task:* **Mapping**
➤ *Variable 1:* **34) SUICIDE**
  ➤ *View:* **Map**

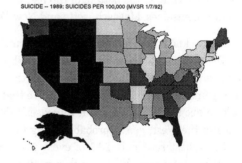

SUICIDE -- 1989: SUICIDES PER 100,000 (MVSR 1/7/92)

To reproduce this graphic on the computer screen using ExplorIt, review the instructions in the *Getting Started* section. For this example, you would open the STATES data file, select the MAPPING task, and select 34) SUICIDE for variable 1. The first view shown is the map view. (Remember, the ➤ symbol indicates which steps you need to perform if you are doing all examples as you follow along in the text. So in the next example, you need only select a new view—that is, you don't need to repeat the first three steps, because they were already done in this example.)

In all the maps we obtain in this workbook, the darker the color of a state (or, when we use the NATIONS data set, the darker the color of a nation), the "more" of a variable it has. Thus, in the map before you, the darker the color, the higher the state's suicide rate. When we look at this map, notice that the "darkest" states—those with the highest suicide rates—tend to be in the West.

The colors indicate which states have higher and lower suicide rates, but do not indicate the actual rate for each state. To find out this rate, you can select the [Legend] button to see what range of rates each color stands for. This still doesn't give us the more precise information we need, so after we obtain a map, we will often obtain a list of each state's actual rate, percent, or value. The ExplorIt Guide to obtain the ranking is:

Data File: **STATES**
Task: **Mapping**
Variable 1: **34) SUICIDE**
➤ View: **List: Rank**

RANK	CASE NAME	VALUE
1	Nevada	23.1
2	Montana	20.0
3	New Mexico	19.5
4	Arizona	18.9
5	Wyoming	17.3
6	Alaska	16.9
7	Oregon	16.7
8	Colorado	16.6
8	Vermont	16.6
10	Florida	16.4

As indicated by the ➤ symbol, if you are continuing from the previous example, select the [List: Rank] option. The number of rows shown on your screen may be different from that shown here. Use the cursor keys and scroll bar to move through the list if necessary.

Inspect the rankings we've obtained. Not surprisingly, Western states appear at the top of the list. Nevada leads the nation in suicide, with a rate of 23.1 per 100,000 population; Montana is just behind with a rate of 20.0. New Mexico, Arizona, Wyoming, Alaska, Oregon, and Colorado come next. The lowest suicide rate, at the bottom of the list, is in New Jersey, where only 6.5 people per 100,000 population kill themselves. Massachusetts and New York rank just above New Jersey.

Now that we know the West has the highest suicide rates, we need to try to explain this regional difference. We are trying to answer the question "Why does the West have the highest regional suicide rate?" If Durkheim is right, it's not enough to say that people in the West are more depressed than those in other regions, and we'd be hard put to prove that anyway. Sociologically speaking, there must be something else about life out west that increases the chances of suicide. Note that we're not saying that if you live in the West, you're probably going to commit suicide. Far from it! Even in Nevada, the state with the highest rate, only 23.1 of every 100,000 people commit suicide. Although that's the highest rate, you're far, far more likely *not* to commit suicide than to commit it if you live in Nevada.

Durkheim thought suicide rates should be higher among groups and in places with less social integration. One indicator of low social integration is what might be called "geographic mobility." The more people move around, the less social integration they have. Let's map a measure of the percent of each state's population not living in the same house 5 years earlier.

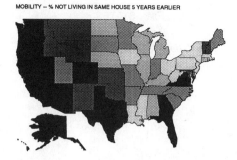

MOBILITY -- % NOT LIVING IN SAME HOUSE 5 YEARS EARLIER

Data File: **STATES**
Task: **Mapping**
➤ Variable 1: **113) MOBILITY**
➤ View: **Map**

If you are continuing from the previous example, return to the variable selection screen and select the variable 113) MOBILITY as the new variable 1. Notice that after selecting a new variable, ExplorIt returns to a map view. (It is not necessary to reselect the STATES data file or the MAPPING task.)

The darker the color, the greater the percent of each state's population that had moved from their dwellings 5 years earlier. The map is not identical to the previous one for suicide, but it is very similar. Generally, the "darker" states—those with the highest geographic mobility and, presumably, the lowest social integration—are also in the West.

Yet another indicator of social integration is population density, or the number of people per square mile. Low population density would indicate low social integration, because it makes sense to think that the fewer people per square mile, the fewer connections they have with one another. Let's map a "lack of population density" variable such that the darker colors indicate the *fewest* people per square mile and the lighter colors the *most* people per square mile.

Data File: **STATES**
Task: **Mapping**
➤ Variable 1: **114) NOT DENSE**
➤ View: **Map**

NOT DENSE -- INVERSE OF POP. PER SQ. MILE

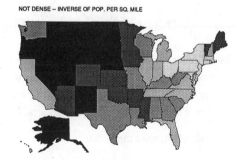

Remember, the darker the color, the lower the population density. Once again, we see a map similar to the suicide map. The West has the lowest population density and also the highest suicide rate.

Let's map one final variable, the percent of each state's population who say they have no religion. All other things being equal, if people don't go to religious services, they have fewer social networks and thus lower social integration. They may also be less likely to have a faith to turn to for comfort in times of personal trouble. All this may help lead to higher suicide rates. If our hypothesis is correct, the percent with no religion should be highest in the West, and in our map the western states should have a darker color.

Data File: **STATES**
Task: **Mapping**
➤ Variable 1: **29) %NO RELIG.**
➤ View: **Map**

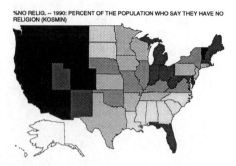

%NO RELIG. -- 1990: PERCENT OF THE POPULATION WHO SAY THEY HAVE NO RELIGION (KOSMIN)

Again, this map looks similar to the suicide one. Lack of religion is highest in the West, the region that also has the highest suicide rate.

Note that none of this *proves* any of Durkheim's theory of suicide. It's possible that suicidal atheists move into the western states and then kill themselves. It's possible that the western states differ in ways other than those mapped here that lead to higher suicide rates. Maybe people in all states are equally likely to try to kill themselves, but those in the West are more skilled at doing so, perhaps because they have more guns. None of this is meant to make light of suicide; rather, we mean to indicate the complexity of explaining social phenomena. Certainly the trends in our maps are consistent with predictions drawn from Durkheim's theory, but they do not prove the theory is correct.

The larger point is that an individual's chances of suicide might depend partly on where the individual lives, on various characteristics of that location, and on other aspects of the individual's social background. It may be true that only unhappy people commit suicide, but an unhappy person in the West is more apt to perform this act than an unhappy one elsewhere.

MAGAZINE CIRCULATION RATES

To take a far less serious topic, let's see if we can explain magazine circulation rates. Some states have higher rates than other states for readership of various magazines. As individuals, we all can decide whether or not to subscribe to a particular magazine, but perhaps our chances of doing so depend on where we live and other aspects of our social backgrounds.

We start with *Gourmet* magazine.

Data File: **STATES**
Task: **Mapping**
➤ Variable 1: **81) GOURMET**
➤ View: **Map**

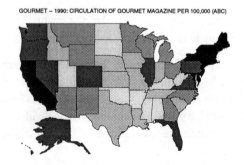

GOURMET -- 1990: CIRCULATION OF GOURMET MAGAZINE PER 100,000 (ABC)

The darker the color, the higher the circulation rate. *Gourmet* circulation rates are generally highest in the Northeast and on the West Coast.

What is it about these regions that promotes interest in this magazine? As the name *Gourmet* implies, perhaps this magazine appeals to wealthy people. Let's map the median family income of each state.

Data File: **STATES**
Task: **Mapping**
➤ Variable 1: **65) MED.FAM $**
➤ View: **Map**

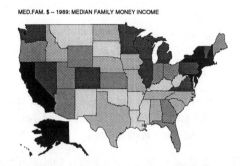

MED.FAM. $ -- 1989: MEDIAN FAMILY MONEY INCOME

Our hypothesis is supported, because the two maps look similar. The Northeast and the West, where *Gourmet* circulation rates are high, are among the wealthiest regions.

Now let's see where *Field & Stream* magazine is most popular. Do you ever read this magazine?

Data File: **STATES**
Task: **Mapping**
➤ Variable 1: **80) F&STREAM**
➤ View: **Map**

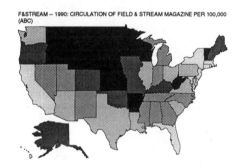

F&STREAM -- 1990: CIRCULATION OF FIELD & STREAM MAGAZINE PER 100,000 (ABC)

Field & Stream circulation is highest in the upper Midwest and mountain states and in northern New England.

Why should this magazine be so popular in these regions of the country? Does living in a rural area promote interest in *Field & Stream*?

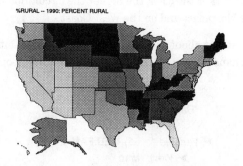

%RURAL -- 1990: PERCENT RURAL

This map appears similar to the last one; the upper Midwest and mountain states and upper New England are among our most rural areas. Not surprisingly, *Field & Stream* magazine does appear most popular in the more rural states.

We'll be dealing with more weighty topics in the rest of the workbook, but it should be clear by now that no matter how much we think we're independent individuals, our choice of magazines depends to some extent on where we live and on certain aspects of our social backgrounds, such as our education and income. If something as trivial as magazine choice and something as serious as suicide both illustrate the sociological perspective, will other aspects of our social lives be different? We'll find out in the exercises ahead.

REVIEW QUESTIONS

Based on the first part of this exercise, answer True or False to the following items:

The sociological perspective emphasizes the influence of our individual personalities on our behavior.	T	F
Durkheim thought that external forces helped to explain suicide rates.	T	F
Suicide rates are higher in the West than in other regions of the United States.	T	F
Population density is higher in the West than in other regions of the United States.	T	F
Field & Stream magazine is most popular in the South.	T	F
The sociological perspective implies that people are totally determined by their social environments.	T	F

EXPLORIT QUESTIONS

You will need to use the ExplorIt software for the remainder of the questions. Make sure you have already gone through the *Getting Started* section that appears before the first exercise. If you have any difficulties using the software to obtain the appropriate information, or if you want to learn additional features of the MAPPING task, refer to Appendix A.

1. Let's examine murder rates in the United States. We will map the number of murders per 100,000 population.

> ➤ *Data File:* **STATES**
> ➤ *Task:* **Mapping**
> ➤ *Variable 1:* **97) MURDER**
> ➤ *View:* **Map**

To select this map using ExplorIt, open the STATES data file, select the MAPPING task, and select 97) MURDER as variable 1.

a. Which of the following regions has the highest murder rate? (Circle one.)   Midwest

New England

South

b. Which of the following regions has the lowest murder rate? (Circle one.)

Midwest

Far West

South

2. Now let's look at the states' actual murder rates.

> Data File: **STATES**
> Task: **Mapping**
> Variable 1: **97) MURDER**
> ➤ View: **List: Rank**

Since you have already selected the appropriate data file, task, and variable, you need only select the [List: Rank] button.

a. Which state has the highest murder rate? _____

b. What is this state's rate? _____

c. Which state has the lowest murder rate? _____

d. What is this state's rate? _____

e. What is the murder rate of the state in which your college or university is located? _____

f. Is this state's murder rate in the top half of the country or in the bottom half?

Top   Bottom

3. What factors might explain the regional differences in murder rates you have just seen? Do you think climate might matter? Let's find out.

> Data File: **STATES**
> Task: **Mapping**
> ➤ Variable 1: **2) WARM WINTR**
> ➤ View: **Map**

The darker the state, the warmer its January temperatures.

a. Which of the following regions is the warmest? (Circle one.)

Midwest

New England

South

  b. Which of the following regions is the least warm? (Circle one.)

Midwest

Far West

South

  c. Does this map look similar to or very different from the murder map?   Similar   Very Different

  d. Do the warmer states tend to have higher murder rates?   Yes   No

  e. What sociological reason might explain the answer you just gave?

4.   What about poverty?

> Data File: **STATES**
> Task: **Mapping**
> ➤ Variable 1: **55) %POOR**
> ➤ View: **Map**

  a. Which region is the poorest? (Circle one.)

Northeast

Midwest

South

West

  b. Is this region the same as the one with the highest murder rate?   Yes   No

5.   While we're looking at poverty, let's determine which states are the poorest and which states are the least poor.

> Data File: **STATES**
> Task: **Mapping**
> Variable 1: **55) %POOR**
> ➤ View: **List: Rank**

  a. List the three poorest states, and list the percent of each state's population that is poor.

STATE	% POOR
_____	_____
_____	_____
_____	_____

b. List the three states with the lowest poverty rates, starting with the one with the lowest rate, and also list the percent of each state's population that is poor.

STATE	% POOR
_____	_____
_____	_____
_____	_____

6. States also differ in their high school dropout rates. Will the states with the highest dropout rates also tend to be the poorest ones?

> Data File: **STATES**
> Task: **Mapping**
> ➤ Variable 1: **76) DROPOUTS**
> ➤ View: **Map**

a. Which region has a higher high school dropout rate? (Circle one.)   Midwest

South

b. Does this map look similar to or very different from the poverty map?   Similar   Very Different

7. Let's see which states have the highest dropout rates and which have the lowest ones.

> Data File: **STATES**
> Task: **Mapping**
> Variable 1: **76) DROPOUTS**
> ➤ View: **List: Rank**

a. List the three states with the highest high school dropout rates, and list the percent of each state's population that dropped out of high school.

STATE	% DROPPED OUT
_____	_____
_____	_____
_____	_____

b. List the three states with the lowest dropout rates, starting with the one with the lowest rate, and also list the percent of each state's population that dropped out.

STATE	% DROPPED OUT
_____	_____
_____	_____
_____	_____

8. Use your answers to the preceding questions to answer True or False to the following items:

The South has the highest regional murder rate in the United States.	T	F
The state with the highest murder rate is New York.	T	F
The region with the highest poverty rate is also the region with the highest murder rate.	T	F
The region with the highest murder rate is also the warmest region.	T	F
Minnesota has the lowest high school dropout rate.	T	F
The poorest region has the highest high school dropout rate.	T	F

9. In the preliminary section of this exercise, you examined regional variations in some magazine subscriptions. Let's look at the subscription rate of yet another magazine, *Cosmopolitan*, which is designed to appeal to young, educated women. We should thus expect *Cosmopolitan* to be especially popular in states with high percentages of college-educated people. Your task here is to determine if there is, in fact, a similarity between states with high *Cosmopolitan* subscription rates and states with high percentages of the college-educated. These are the steps you should follow:

a. Create a map of 82) COSMO.

b. Obtain a ranked list of 82) COSMO. Print out this ranked distribution and turn it in with your assignment.

c. Next, create a map of 75) COLL.DEGR

d. Obtain a ranked list of 75) COLL.DEGR. Print out this ranked distribution and turn it in with your assignment.

e. As your final step, compare the maps for both variables and determine whether they look similar. In particular, determine whether the regions of the U.S. with high *Cosmopolitan* subscription rates are also those with high percentages of the college educated, and whether the regions with low subscription rates are also those with low percentages of the college-educated. Next, compare the rankings you printed out for both variables. Determine whether the states that are high on one list tend to be high on the other list, and whether the states that are low on one list tend to be low on the other list. As one way of doing this, fill in the ranks for each of the following states:

	75) COLL.DEGR	82) COSMO
MASSACHUSETTS	_____	_____
CONNECTICUT	_____	_____
NEW JERSEY	_____	_____
NEW HAMPSHIRE	_____	_____
WEST VIRGINIA	_____	_____
ARKANSAS	_____	_____
KENTUCKY	_____	_____
MISSISSIPPI	_____	_____

f. What conclusion do you draw regarding the areas in which *Cosmopolitan* is most popular?

CULTURE AND SOCIETY

Tasks: Mapping, Scatterplot, Univariate
Data Files: NATIONS, STATES, GSS

E very society has its own culture, which is both material and nonmaterial. *Material culture* refers to the tangible objects that make up every society and includes such things as eating utensils, clothing, and artwork. Societies differ in the types of material culture they contain. *Nonmaterial culture* refers to the language, symbols, values, beliefs, and behaviors that characterize a society and that the society's members learn through socialization. If the culture we learn influences our attitudes, behaviors, and values, culture is a key concept to the sociological perspective.

This chapter sketches some key cultural differences throughout the world and within the United States. As you develop your appreciation of culture's importance, think of the many ways in which you've been influenced by the culture in which you've grown up.

CULTURE AROUND THE WORLD

Let's start with one aspect of a society's material culture, the car, and see how nations differ in car ownership.

> ➤ *Data File:* **NATIONS**
> ➤ *Task:* **Mapping**
> ➤ *Variable 1:* **34) CARS/1000**
> ➤ *View:* **Map**

CARS/1000 – CARS PER 1000 PERSONS (WABF, 1995)

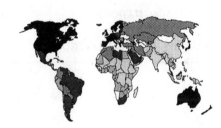

> To reproduce this graphic on the computer screen using ExplorIt, review the instructions in the *Getting Started* section. For this example, you would open the NATIONS data file, select the MAPPING task, and select 34) CARS/1000 for variable 1. The first view shown is the Map view. (Remember, the ➤ symbol indicates which steps you need to perform if you are doing all examples as you follow along in the text. So in the example that follows, you need only select a new view—that is, you don't need to repeat the first three steps, because they were already done in this example.)

Remember, the darker the color, the more cars per 1000 population. Generally, North America and Europe have the most cars, and Africa and Asia the least.

Data File: **NATIONS**
Task: **Mapping**
Variable 1: **34) CARS/1000**
➤ View: **List: Rank**

RANK	CASE NAME	VALUE
1	United States	554.0
2	Luxemburg	502.2
3	Italy	486.0
4	Canada	460.9
5	Iceland	459.3
6	New Zealand	445.3
7	Germany	439.5
8	Australia	431.9
9	Switzerland	429.4
10	Brunei	415.2

The United States leads the world, with 554 cars for every 1000 people, while Luxemburg, Italy, Canada, and Iceland rank 2 through 5. At the other extreme are nations with hardly any cars. Cambodia ranks last, with only 0.4 car per 1000 people. Four other nations—Rwanda, Myanmar, Ethiopia, and Bangladesh—all have 1.0 car or less per 1000 people.

Now let's examine nonmaterial culture. We'll start with an important behavior, attending religious services.

The World Values Survey (WVS) data that are part of the NATIONS data set include a measure of the percent of each nation's population who attend religious services at least monthly. This time let's go directly to the ranked list of nations.

Data File: **NATIONS**
Task: **Mapping**
➤ Variable 1: **51) CH.ATTEND**
➤ View: **List: Rank**

RANK	CASE NAME	VALUE
1	Ireland	88.00
2	Nigeria	87.00
3	Poland	85.00
4	India	76.00
5	South Korea	64.00
6	Mexico	63.00
7	United States	57.00
8	Argentina	55.00
9	Italy	51.00
10	Brazil	50.00

In Ireland, 88 percent of the public attends religious services at least monthly for the highest rate in the data set. Nigeria and Poland rank just behind, however, at 87 and 85 percent respectively. Attendance in the United States is a bit lower, at 57 percent. At the other end, only 1 percent of China's population attends religious services at least monthly, and only 6 percent of people in Russia and Belarus (another former Soviet republic) do so. Religion is clearly more a part of some nations' cultures than of other nations' cultures.

Since we just viewed the rankings for religious attendance, let's look at the map for a variable for the percent of residents who say God is important in their lives.

Data File: **NATIONS**
Task: **Mapping**
➤ *Variable 1:* **52) GOD IMPORT**
➤ *View:* **Map**

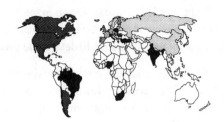

We'd expect this map to look similar to the religious attendance one, and it does. Europe again ranks fairly low on this measure of religious belief; several other nations report much higher percentages of belief in God's importance. Once again we see that religion plays a more important role in some nations' cultures than in other nations' cultures.

We can compare the maps for religious attendance and God's importance a bit more quickly with the following ExplorIt Guide.

Data File: **NATIONS**
Task: **Mapping**
➤ *Variable 1:* **51) CH.ATTEND**
➤ *Variable 2:* **52) GOD IMPORT**
➤ *Views:* **Map**

CH.ATTEND -- PERCENT WHO ATTEND RELIGIOUS SERVICES ONCE A MONTH OR MORE (WVS)

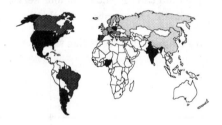

r = 0.819**

GOD IMPORT -- PERCENT SAYING GOD IS IMPORTANT IN THEIR LIVES (WVS)

> **If you are continuing from the previous example, return to the variable selection screen for the MAPPING task. Select 51) CH.ATTEND for variable 1 and 52) GOD IMPORT for variable 2.**

Now we again see the two maps we've already seen, but both at once.

It certainly makes sense to think that the countries where people are most likely to say God is important in their lives are also the countries where religious attendance is highest, which is what the two maps indicate. But how similar do two maps have to appear for us to say that they're similar? How different do they have to appear for us to conclude that they're different? When we just look at the maps visually, it's sometimes difficult to determine precisely how similar or different they are.

Fortunately, there's a simple method for determining how similar or different any two maps are. This method is called the ***scatterplot*** and was invented a century ago by English scholar Karl Pearson. You may have encountered a scatterplot when you learned algebra, but if you didn't, or if you don't remember whether you did, don't let that worry you. It's easy to understand.

To develop a scatterplot, we first draw a horizontal line to represent the map of God's importance. Recall that China ranked lowest on this variable, with only 7 percent of its population saying God is important in their lives. At the left end of our horizontal line, we write 7 to indicate China. At the right end, we write 99 to represent Nigeria, which we saw had the highest percent, 99, citing God's importance. Our horizontal line looks like this:

7   99

To represent the map of religious attendance, we now draw a vertical line up the left side of the horizontal line (to form an "L"). At the bottom of the new line, we write 1 to represent the percent of China's population that attends religious services at least monthly, since China had the lowest percent. At the top, we write 88 to represent Ireland's percent, which ranked highest.

We now have a figure that looks like this:

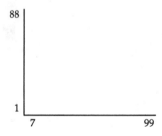

Now that we have a line with an appropriate scale to represent each map, we next look back at the rankings for each map to learn the value for each nation and then locate it on each line according to its score. Let's start with Nigeria. For the God's importance variable, Nigeria had the highest percent, 99, so we can easily find its place on the horizontal line. Make a small mark at 99 to locate Nigeria. For the religious attendance variable, Nigeria ranked second at 87 percent, so we make a mark on the vertical line just under the 88 that's already there (for Ireland). Next we draw a line up from the mark for Nigeria on the horizontal line and draw another line to the right from the mark for Nigeria on the vertical line. Where these two lines intersect, we place a dot. This dot represents the *combined* map locations of Nigeria.

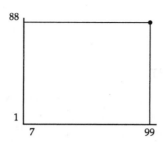

Now let's find the United States. Its value for the God's importance variable was 88, so we estimate where 88 is on the horizontal line and make a mark at that spot. The U.S. percentage for the religious attendance variable was 57, so we make a mark on the vertical line where 57 would be. We next draw a line up from the mark on the horizontal line and one out to the right from the mark on the vertical line. The place where these lines meet is the combined map location for the United States. It's below and to the left of that for Nigeria.

When we have followed this procedure for each nation, we will have 35 dots, one for each nation, located within the space defined by the vertical and horizontal lines representing the two maps. We will have created a scatterplot. Fortunately, ExplorIt does all the work for you with just a few mouse clicks. The ExplorIt Guide that follows shows you how it's done.

Data File: **NATIONS**	
➤ *Task:* **Scatterplot**	
➤ *Dependent Variable:* **51) CH.ATTEND**	
➤ *Independent Variable:* **52) GOD IMPORT**	

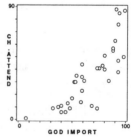

$r = 0.819^{**}$ Prob. = 0.000   N = 35   Missing = 139

Notice that the scatterplot requires two variables.

Special Feature: When the scatterplot is showing, you may obtain the information on any dot by clicking on it. A little box will appear around the dot, and the values of 51) CH.ATTEND (or the x-axis variable) and of 52) GOD IMPORT (or the y-axis variable) will be shown.

Each of these dots represents a nation. We can see that the nations with the highest percentages on the God's importance variable (the ones to the right of the scatterplot) also tend to be the nations with the highest percentages on the religious attendance variable (the ones nearer to the top of the scatterplot).

Once Pearson had created the scatterplot, his next step was to calculate what he called the ***regression line***.

Data File:	**NATIONS**
Task:	**Scatterplot**
Dependent Variable:	**51) CH.ATTEND**
Independent Variable:	**52) GOD IMPORT**
➤ Display:	**Reg. Line**

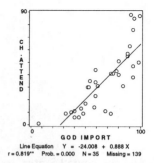

To show the regression line, select the [Reg. Line] option from the menu.

The regression line represents the line that best summarizes the trend of the dots; it's the line that comes closest to connecting all the dots. The closer the dots in any scatterplot are to the regression line, the more alike are the two maps represented in the scatterplot. The farther the dots are from the regression line and, thus, the less they resemble a straight line, the less the variables represented by the two maps have anything to do with each other.

To make it simpler to interpret the dots in a scatterplot, Pearson invented a statistic called the ***correlation coefficient*** and used the letter *r* as the symbol for this coefficient. The correlation coefficient varies from 0.0 to 1.0. When two maps are identical (as when you have two maps of the same variable), the correlation coefficient will be 1.0. When the maps are completely different from each other, the coefficient will be 0.0. The closer the correlation coefficient is to 1.0, then, the more alike are the two maps.

Look at the lower part of the screen above and you will see r = 0.819. This indicates that the maps are very similar. Ignore the asterisks after this figure for now.

Correlation coefficients can be positive or negative. The one we've just seen is positive: when the God's importance percent is higher, the religious attendance percent is higher. As one variable goes up, so does the other. But negative correlations are also possible.

Data File:	**NATIONS**
Task:	**Scatterplot**
➤ Dependent Variable:	**9) FERTILITY**
➤ Independent Variable:	**40) EDUCATION**
➤ Display:	**Reg. Line**

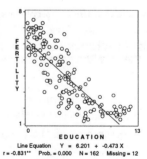

Our dependent variable is the average number of children born to a woman in her lifetime, and our independent variable is the average number of years of school a nation's adults have attended. Not surprisingly, the more educated a nation's population, the lower its fertility rate. Notice that the regression line now slopes downward from left to right, rather than upward, as in the previous example. This type of slope always indicates a negative correlation. Thus, you'll also notice that a minus sign now precedes the correlation coefficient: $r = -0.831$.

Some correlations are above zero but are still too small for us to conclude that two variables are related. When this is the case, we treat the correlations as if they were random accidents and conclude that the correlation is in effect zero. The software automatically tells you whether the correlation was an accident or, instead, whether it was most likely due to a real relationship between the variables. If you look back at the correlation between God's importance and religious attendance, you'll see that two asterisks follow the value of r ($r = 0.819**$). Two asterisks means there is less than 1 chance in 100 that this correlation is a random accident. One asterisk means there is less than 1 chance in 20 that it's an accident. When no asterisks follow a correlation coefficient, the chances are too high that it could be a random accident, and we conclude there is no correlation at all. *Treat all correlations without asterisks as zero correlations*.

When r has at least one asterisk, it indicates a statistically significant relationship. In assessing the strength of this correlation, treat an r smaller than .3 (in absolute value) as a weak relationship, an r between .3 and .6 as a moderate relationship, and an r greater than .6 as a strong relationship.

Keep in mind, though, that correlation does not necessarily mean causation. Sometimes two variables can be correlated without one affecting the other. For example, in the months when more ice cream is sold, the crime rate is higher. Does that mean eating ice cream causes crime? Does it mean that fear of crime causes people to eat ice cream? Of course not! Obviously, more ice cream is sold in the summer months, and, for quite different reasons, crime is higher during the summer months. As another example, the more people listen to rock music, the worse their acne is. Does that mean that listening to rock music excites your pores and causes you to break out? Could it mean that having bad acne forces you to stay home and listen to rock to make you feel better? Not at all! Obviously, younger people have worse acne, and younger people, for different reasons, are more apt to listen to rock music. A correlation between two variables might indicate a cause-and-effect relationship, but it doesn't have to.

A correlation between two variables also doesn't automatically tell us which variable is affecting which. This is called the "causal order" or "chicken and egg" question. In the religion example earlier, we saw a strong relationship between God's importance and religious attendance. But it's not clear which variable is causing which. Are people going to religious services more often because they believe God is important in their lives, or do they believe God is important in their lives because they're going to religious services more often? Or could both "causal directions" be valid?

CULTURE IN THE UNITED STATES

A few maps will indicate the regional distribution of aspects of the U.S. material culture. We'll start with the percent of occupied housing structures lacking a telephone.

➤ *Data File:* **STATES**
    ➤ *Task:* **Mapping**
➤ *Variable 1:* **48) NO PHONES**
    ➤ *View:* **Map**

NO PHONES -- 1990: PERCENT OF OCCUPIED HOUSING UNITS WITH NO TELEPHONE

Notice that the STATES data file must be open.

The darker the color, the greater the percent of a state's households without a telephone. The South has the highest rate of households without a telephone. What accounts for regional variation in the presence of an essential object in the U.S. material culture?

In some parts of the country, the pickup truck is very much a part of the culture of a local area or even that of an entire state. Let's see where the pickups are.

    Data File: **STATES**
       Task: **Mapping**
➤ *Variable 1:* **105) PICKUPS**
    ➤ *View:* **Map**

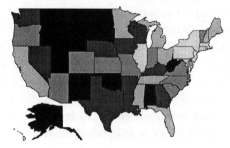

PICKUPS -- 1993: PICKUP TRUCKS PER 1000 POPULATION (HIGHWAY)

There seem to be more pickups west of the Mississippi River, and especially in the northern part of that region of the country.

    Data File: **STATES**
       Task: **Mapping**
  Variable 1: **105) PICKUPS**
    ➤ *View:* **List: Rank**

RANK	CASE NAME	VALUE
1	Wyoming	548.3
2	Montana	398.3
3	South Dakota	368.7
4	Idaho	312.5
5	North Dakota	304.3
6	New Mexico	304.2
7	Alaska	292.3
8	Alabama	274.0
9	Oklahoma	273.6
10	Utah	259.9

Wyoming has 548.3 pickups per 1000 people, quite a high number, but Connecticut has only 30.0 per 1000 people.

Discovering Sociology

Earlier we asked what accounts for the regional variation in the lack of telephones in the United States. It makes sense to think that poorer states will be more likely to lack phones. Let's check this out with a scatterplot.

> Data File: **STATES**
> ➤ Task: **Scatterplot**
> ➤ Dependent Variable: **48) NO PHONES**
> ➤ Independent Variable: **55) %POOR**
> ➤ Display: **Reg. Line**

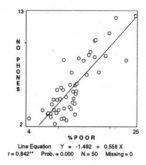

Line Equation   Y = -1.492 + 0.558 X
r = 0.842**   Prob. = 0.000   N = 50   Missing = 0

The poorer a state, the higher its rate of non-telephone households. The correlation coefficient, r, is a very high 0.84**. (We have rounded r to two digits after the decimal point and will do that from now on.)

We can also gauge the U.S. culture with survey questions given to a national, random sample of the adult, non-institutionalized U.S. population. The General Social Survey (GSS) is one such survey and has been administered regularly since 1972; this workbook uses the 1996 version. A random sample is just what it sounds like. As a sample, it's a much smaller subset of the entire population; the number of people in the 1966 GSS sample is 2904. It's random because everyone in the country has the same chance of being in the sample. Although fewer than 3000 end up in the sample, we all had the same chance of being included. Random samples save a lot of time and money, because we can gauge attitudes and behaviors of the entire population, well over 200 million, with under 3000 respondents. Because the sample is random, we can generalize the results from the sample to the entire population. More on that in the next chapter.

One value often associated with the United States is the belief in the importance of democracy. The GSS asked its respondents how proud they are of the way democracy works.

> ➤ Data File: **GSS**
> ➤ Task: **Univariate**
> ➤ Primary Variable: **111) PROUD DEM**
> ➤ View: **Pie**

PROUD DEM -- HOW PROUD IS RESPONDENT OF...The way democracy works
(PROUDDEM)

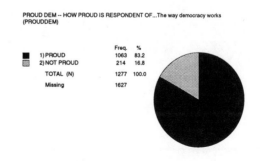

		Freq.	%
■	1) PROUD	1063	83.2
▒	2) NOT PROUD	214	16.8
	TOTAL (N)	1277	100.0
	Missing	1627	

Notice that the GSS data file must be open.

About 83 percent of the sample say they're proud of the way democracy works, and about 17 percent say they're not proud. Clearly a belief in democracy is part of the U.S. culture.

Another value in the U.S. culture is independence, which the country has valued ever since colonial days. The GSS asks, "In general, would you say that people should obey the law without exception, or are there exceptional occasions on which people should follow their consciences even if it means breaking the law?"

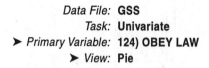

Data File: **GSS**
Task: **Univariate**
➤ *Primary Variable:* **124) OBEY LAW**
➤ *View:* **Pie**

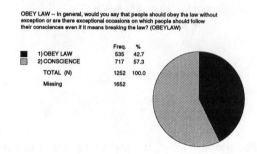

OBEY LAW -- In general, would you say that people should obey the law without exception or are there exceptional occasions on which people should follow their consciences even if it means breaking the law? (OBEYLAW)

		Freq.	%
■	1) OBEY LAW	535	42.7
▨	2) CONSCIENCE	717	57.3
	TOTAL (N)	1252	100.0
	Missing	1652	

More than half of the sample, or 57.3 percent, think people should obey their consciences even if it means breaking the law. The spirit of the colonial period still lives in America.

EXERCISE

2

REVIEW QUESTIONS

Based on the first part of this exercise, answer True or False to the following items:

Telephones and television are examples of nonmaterial culture.	T	F
Religious attendance is higher in Europe than in North America.	T	F
In the United States, households in the Midwest are especially likely not to have phones.	T	F
In the United States, the poorer a state, the higher the rate of households without telephones.	T	F
Judging from the GSS, less than half of Americans believe that we should obey our consciences even if it means breaking the law.	T	F
A nonzero correlation always tells us that one variable is affecting another variable.	T	F

EXPLORIT QUESTIONS

> **If you have any difficulties using the software to obtain the appropriate information, or if you want to learn additional features of the MAPPING or SCATTERPLOT tasks, refer to Appendix A.**

1. People of different nationalities living in the same society sometimes have different subcultures. Let's see which states have the highest percentages of people with Japanese ancestry.

> ➤ *Data File:* **STATES**
> ➤ *Task:* **Mapping**
> ➤ *Variable 1:* **22) %JAPANESE**
> ➤ *View:* **Map**

   a. Which region of the country has the highest percentage of residents with Japanese ancestry? (Circle one.)

   Northeast

   South

   Midwest

   West

b. Which region has the lowest percentage? (Circle one.)   Northeast

South

West

c. What might account for the regional patterns that you see?

2. To continue our look at Japanese ancestry, let's look at each state's ranking.

> *Data File:* **STATES**
> *Task:* **Mapping**
> *Variable 1:* **22) %JAPANESE**
> ➤ *View:* **List: Rank**

a. Which state has the highest percent with Japanese ancestry?   _____

b. What is its percent?   _____%

c. Which state has the lowest percent with Japanese ancestry?   _____

d. What is its percent?   _____%

3. Now we'll look at the states in terms of their proportions of college graduates and their median family incomes.

> *Data File:* **STATES**
> *Task:* **Mapping**
> ➤ *Variable 1:* **75) COLL.DEGR.**
> ➤ *Variable 2:* **65) MED.FAM. $**
> ➤ *Views:* **Map**

a. These maps are (circle one):   Almost identical

Similar, but it is somewhat
difficult to determine how similar

Not very similar

Almost opposite

b. Which region generally has the highest proportion of college
    graduates? (Circle one.)

                                                                Northeast

South

Midwest

West

c. Which region has the lowest proportion overall? (Circle one.)

Northeast

South

Midwest

West

d. Which region generally has the highest median family
    income? (Circle one.)

Northeast

South

Midwest

West

e. Which region has the lowest income overall? (Circle one.)

Northeast

South

Midwest

West

4. Now we'll compare these maps by using a scatterplot.

> Data File: **STATES**
> ➤ Task: **Scatterplot**
> ➤ Dependent Variable: **75) COLL.DEGR.**
> ➤ Independent Variable: **65) MED.FAM. $**
> ➤ Display: **Reg. Line**

a. The states that are highest on 75) COLL.DEGR. should appear
    as dots at the (circle one):

Right of the scatterplot

Left of the scatterplot

Top of the scatterplot

Bottom of the scatterplot

b. The dots appearing at the bottom of the scatterplot represent the states that have the (circle one):

Highest median incomes

Lowest median incomes

Highest percent of college graduates

Lowest percent of college graduates

c. What is the value of r for this scatterplot?

r = _____

d. Is r statistically significant?

Yes   No

e. The scatterplot indicates that states with higher family incomes (circle one):

Have higher percents of college graduates

Have lower percents of college graduates

Have about the same percents of college graduates as states with low family incomes

f. Can you conclude from the scatterplot that income affects the chances of graduating from college? Why or why not?

5. Recall that we examined the distribution of telephone ownership in the United States. Let's now look at phone ownership around the world, using a measure of the number of phones per 1000 population.

> *Data File:* **NATIONS**
> *Task:* **Mapping**
> *Variable 1:* **37) PHONE 1000**
> *Display:* **Map**

a. Overall, which of the following areas has the highest rate of telephone ownership? (Circle one.)

North America

South America

Africa

b. Which has the lowest rate of phone ownership? (Circle one.)

North America

South America

Africa

6. To continue our exploration, let's see how the various nations rank on this variable.

> *Data File:* **NATIONS**
> *Task:* **Mapping**
> *Variable 1:* **37) PHONE 1000**
> ➤ *View:* **List: Rank**

a. Which country has the highest rate of telephone ownership?   _____

b. What is its rate?   _____

c. Which nation has the lowest rate of telephone ownership?   _____

d. What is this nation's rate?   _____

e. Does the United States rank in the top ten?   Yes   No

7. Recall that the poorest states in the United States had the lowest rates of phone ownership. Let's hypothesize that we'll find a similar link between poverty and phone ownership around the world. Our independent variable will be the annual national product per capita, a rough measure of a nation's wealth or poverty.

> *Data File:* **NATIONS**
> ➤ *Task:* **Scatterplot**
> ➤ *Dependent Variable:* **37) PHONE 1000**
> ➤ *Independent Variable:* **29) $ PER CAP**
> ➤ *Display:* **Reg. Line**

Examine this scatterplot and note its correlation coefficient. In the space below, write a paragraph stating the hypothesis we're testing, reporting the results of your analysis, and drawing a conclusion about whether the data support this hypothesis.

8. The work ethic is a value that is often said to be an important part of the American nonmaterial culture. The GSS asks whether people get ahead by their own hard work or, instead, by luck or help from others. If the work ethic is that important, a majority of the public should say hard work is how people get ahead. Let's see if this is true.

> *Data File:* **GSS**
> *Task:* **Univariate**
> *Primary Variable:* **149) GET AHEAD?**
> *View:* **Pie**

a. What percent say people get ahead through hard work? _____%

b. What percent say people get ahead by being lucky? _____%

c. Do these results support the belief that the work ethic is an important value in American society? Why or why not?

9. Related to the work ethic is the belief that American society allows people to improve their lot in life if they work hard enough. The GSS asks whether respondents agree that "people like me have a good chance of improving our standard of living." Let's see whether a majority of the public agrees with this statement.

Data File: **GSS**
Task: **Univariate**
> *Primary Variable:* **145) GOOD LIFE**
> *View:* **Pie**

a. What percent of the sample agrees that they have a good chance of improving their standard of living? _____%

b. Did you think this percent would be higher than it was, lower than it was, or about what you found? Explain your answer.

c. Using your results for the GOOD LIFE and GET AHEAD? variables, how would you characterize the American culture?

d. In your opinion, how realistic are the views expressed through these two variables? Is life pretty much like what the majority of the public thinks?

SOCIALIZATION

Tasks: Mapping, Scatterplot, Univariate, Cross-tabulation, Auto-Analyzer
Data Files: NATIONS, GSS

I f people are to become social beings and not just individuals, they must first learn what their society expects of them. They must learn their society's norms, values, and symbols, and other aspects of their society's culture. Socialization is the process by which people learn all these things and more. Although socialization begins in infancy and is perhaps most important during the many years before adulthood, it continues throughout our mature years as well. A society without socialization is impossible.

This chapter looks at some aspects of socialization around the world and within the United States. Our international focus will be on values that people in different nations think children should learn. We'll see that the importance people place on these values differs among nations and is related to ways in which nations differ. We'll also look at these values with the GSS sample and use other variables to illustrate how our parents influence our thinking and behavior.

INTERNATIONAL DIFFERENCES IN SOCIALIZATION

Several countries in the NATIONS data set were asked whether it's important that children learn four different values: (1) good manners; (2) independence; (3) obedience; and (4) thriftiness. Let's map each of these.

> *Data File:* **NATIONS**
> *Task:* **Mapping**
> *Variable 1:* **129) KID MANNER**
> *View:* **Map**

KID MANNER -- PERCENT WHO THINK IT IS IMPORTANT THAT CHILDREN LEARN GOOD MANNERS (WVS)

The importance placed on children's learning good manners certainly varies around the world, and no clear geographic pattern really emerges.

Data File: **NATIONS**
Task: **Mapping**
➤ *Variable 1:* **130) KID INDEPN**
➤ *View:* **Map**

KID INDEPN -- PERCENT WHO THINK IT IS VERY IMPORTANT THAT A CHILD DEVELOP INDEPENDENCE (WVS)

Again, not the clearest pattern, but China and several nations in Western Europe seem most likely to think that children need to develop independence.

Data File: **NATIONS**
Task: **Mapping**
➤ *Variable 1:* **131) KID OBEY**
➤ *View:* **Map**

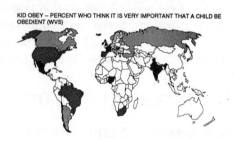

KID OBEY -- PERCENT WHO THINK IT IS VERY IMPORTANT THAT A CHILD BE OBEDIENT (WVS)

Several European nations rank rather low on this variable, but there is not a clear pattern overall.

The final socialization variable is the percent who say it's very important that a child learn to be thrifty.

Data File: **NATIONS**
Task: **Mapping**
➤ *Variable 1:* **132) KID THRIFT**
➤ *View:* **Map**

KID THRIFT -- PERCENT WHO THINK IT IS VERY IMPORTANT THAT A CHILD LEARN THRIFT (WVS)

The highest percentages here seem to be found in Europe and parts of Asia.

What explains these international differences in the perceived importance of different values for children's socialization? As just one example, do you think that when people are more educated, they should be more likely to value independence? If so, the level of a nation's education should be related to the importance its citizens place on independence. Let's find out.

<table>
<tr><td align="right">Data File:</td><td>**NATIONS**</td></tr>
<tr><td align="right">➤ Task:</td><td>**Scatterplot**</td></tr>
<tr><td align="right">➤ Dependent Variable:</td><td>**130) KID INDEPN**</td></tr>
<tr><td align="right">➤ Independent Variable:</td><td>**40) EDUCATION**</td></tr>
<tr><td align="right">➤ Display:</td><td>**Reg. Line**</td></tr>
</table>

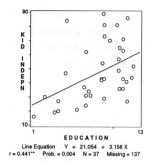

The more educated a nation's population, the more likely they are to say that it's very important for children to learn independence (r = .44**). Our hypothesis is supported.

SOCIALIZATION IN THE UNITED STATES

GSS respondents were asked to rank the importance of five qualities for children to learn. These five qualities were: (1) to think for themselves; (2) to obey; (3) to work hard; (4) to help others when they need help; and (5) to be well-liked or popular. Your GSS data set includes a variable giving the percent of respondents who ranked a particular quality *first* in importance. Which quality would you rank first? Compare your answer to the GSS result.

<table>
<tr><td align="right">➤ Data File:</td><td>**GSS**</td></tr>
<tr><td align="right">➤ Task:</td><td>**Univariate**</td></tr>
<tr><td align="right">➤ Primary Variable:</td><td>**95) KID LEARN**</td></tr>
<tr><td align="right">➤ View:</td><td>**Pie**</td></tr>
</table>

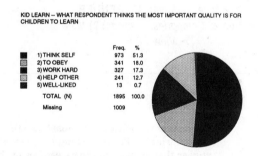

KID LEARN -- WHAT RESPONDENT THINKS THE MOST IMPORTANT QUALITY IS FOR CHILDREN TO LEARN

		Freq.	%
■	1) THINK SELF	973	51.3
▨	2) TO OBEY	341	18.0
▨	3) WORK HARD	327	17.3
▨	4) HELP OTHER	241	12.7
■	5) WELL-LIKED	13	0.7
	TOTAL (N)	1895	100.0
	Missing	1009	

About 51 percent of GSS respondents ranked thinking for oneself as first in importance for children to learn. Eighteen percent ranked obeying as first in importance, and about 17 percent ranked working hard as the most important quality. Almost 13 percent ranked helping others as most important, and less than 1 percent said being well-liked was most important.

Let's focus on two of these values, thinking for oneself and learning to obey, and ask what social factors influence which of these we think is more important. To do this, we'll use another variable that includes only those respondents who ranked thinking for oneself or learning to obey first in importance. Here's the distribution of this variable.

  Data File: GSS
  Task: Univariate
  ➤ <i>Primary Variable:</i> 96) THINK/OBEY
  ➤ <i>View:</i> Pie

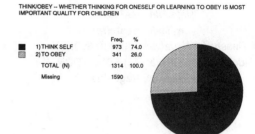

THINK/OBEY -- WHETHER THINKING FOR ONESELF OR LEARNING TO OBEY IS MOST IMPORTANT QUALITY FOR CHILDREN

		Freq.	%
■	1) THINK SELF	973	74.0
▨	2) TO OBEY	341	26.0
	TOTAL (N)	1314	100.0
	Missing	1590	

Almost three-quarters of these respondents ranked thinking for oneself first in importance for children, and about one-fourth ranked learning to obey first in importance.

Do you think gender will make a difference in which of these two qualities people favor? We often hear that men are more authoritarian than women and more the "disciplinarian" in the home with children. If so, men should be more likely than women to think that children should obey. Let's find out. We'll first obtain the distribution of our "thinking/obeying" variable just for men. To do so, we'll use ExplorIt's subset variable capability.

  Data File: GSS
  Task: Univariate
  Primary Variable: 96) THINK/OBEY
  ➤ <i>Subset Variable:</i> 45) GENDER
  ➤ <i>Subset Categories:</i> Include: 2) Male
  ➤ <i>View:</i> Pie

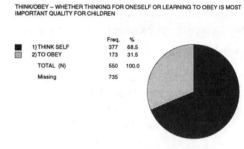

THINK/OBEY -- WHETHER THINKING FOR ONESELF OR LEARNING TO OBEY IS MOST IMPORTANT QUALITY FOR CHILDREN

		Freq.	%
■	1) THINK SELF	377	68.5
▨	2) TO OBEY	173	31.5
	TOTAL (N)	550	100.0
	Missing	735	

[Subset]

The option for selecting a subset variable is located on the same screen you use to select other variables. For this example, select 45) GENDER as a subset variable. A window will appear that shows you the categories of the subset variable. Select 2) Male as your subset category and choose the [Include] option. Then click [OK] and continue as usual.

With this particular subset selected, the results will be limited to the males in the sample. The subset selection continues until you exit the task, delete all subset variables, or clear all variables.

When we look only at men, we see that 31.5 percent rank obedience as the most important quality for children, and 68.5 percent rank thinking for oneself as most important. Now let's look only at women.

Data File:	**GSS**	
Task:	**Univariate**	
Primary Variable:	**96) THINK/OBEY**	
➤ *Subset Variable:*	**45) GENDER**	
➤ *Subset Categories:*	**Include: 1) Female**	
➤ *View:*	**Pie**	

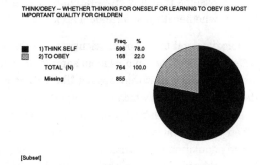

THINK/OBEY -- WHETHER THINKING FOR ONESELF OR LEARNING TO OBEY IS MOST
IMPORTANT QUALITY FOR CHILDREN

	Freq.	%
■ 1) THINK SELF	596	78.0
▨ 2) TO OBEY	168	22.0
TOTAL (N)	764	100.0
Missing	855	

[Subset]

> The easiest way to change the subset category to Females (from Males) is to first delete
> the subset variable 45) GENDER. Then reselect 45) GENDER as the subset variable.
> Include 1) Female as your subset category. Then click [OK] and continue as usual.

When we look just at women, we find that only 22.0 percent rank obedience as the most important quality for children, and 78.0 percent rank thinking for oneself as most important. If we subtract the 22.0 percent for women from the 31.5 percent for men, we see that there is a 9.5 percent point difference and that men are more likely than women to rank obedience as the most important quality for children.

We found this gender difference using a two-step process that involves obtaining two univariate distributions. Your software includes a one-step procedure for obtaining the same result. This procedure is called a ***cross-tabulation*** and yields a table including the same data you saw in the two univariate distributions. Here's how it looks.

Data File:	**GSS**	
➤ *Task:*	**Cross-tabulation**	
➤ *Row Variable:*	**96) THINK/OBEY**	
➤ *Column Variable:*	**45) GENDER**	
➤ *View:*	**Tables**	
➤ *Display:*	**Column %**	

	45) GENDER	
	FEMALE	MALE
96) THINK SELF	78.0%	68.5%
TO OBEY	22.0%	31.5%
TOTAL	100.0%	100.0%

V=0.107**

> To construct this table, return to the main menu and select the CROSS-TABULATION task.
> Then select 96) THINK/OBEY as the row variable and 45) GENDER as the column
> variable. When the table is showing, select the [Column %] option.

The table indicates that 22.0 percent of women rank obedience first in importance, compared to 31.5 percent of men. This matches the univariate results obtained above, as it should.

Remember that the GSS is a sample, or a subset of the larger U.S. population. Does this 9.5 percent difference reflect an actual difference in the entire population, or is it something we got just by accident? If we flipped 1000 coins, you know that we "should" get 500 heads, but you also know that someone could easily get 482 heads and someone else 510 heads. By the same token, a sample can sometimes differ from a population for random reasons. Because of this, we need to determine whether we can assume that any cross-tabulation's result does indeed reflect what's happening in the population, or whether it's just a random accident.

Sociologists and other social scientists use tests of statistical significance to make this determination. Differences observed in a random sample are said to be statistically significant when these differences are high enough that they are not likely to be a random accident, and thus reflect a real difference in the population from which the sample is drawn. In sociology, we assume that any observed difference in a sample is statistically significant when it occurs fewer than 5 times out of 100 by chance alone. When we say that a difference in a sample is statistically significant, we conclude that it really exists in the population from which the sample is drawn. If a difference would occur more than 5 times out of 100 by chance alone (a probability level of .05), we say that the difference is not statistically significant, and we conclude that it does not exist in the population from which the sample is drawn.

Some sociologists think a .05 probability level is too high and instead prefer a .01 level. That is, they assume that a difference is statistically significant only if it would occur less than 1 time out of 100 by chance alone.

To see what the level of significance is for the table we calculated above, look near the bottom where it reads "V = .107." The letter V stands for Cramer's V, which is a statistic indicating the strength of a relationship in cross-tabulations. Consider it the equivalent of Pearson's r, which you used in the previous exercise. If V is followed by one asterisk, the difference it represents is statistically significant at the .05 level of probability. That is, it would be expected to occur fewer than 5 times out of 100 by chance alone. If V is followed by two asterisks, the difference it represents is statistically significant at the .01 level of probability; it would be expected to occur less than 1 time out of 100 by chance alone. Whether V has one or two asterisks after it, we conclude the difference in the sample is statistically significant and thus exists in the entire U.S. population. If V is followed by *no* asterisks, the difference it represents is not statistically significant, and we conclude that there is no difference in the U.S. population.

Notice that the V of .107 in our table is followed by two asterisks. This means that the gender difference in the table would occur by chance alone less than 1 time out of 100. We can conclude with high confidence that the gender difference exists in the entire U.S. population.

Sometimes in large samples such as the GSS, you'll see that a difference of only 2–3 percent can be statistically significant. This difference is small and, in fact, "nothing to write home about," even if it exists in the entire population. As this example indicates, a difference that is statistically significant might not be *substantively significant*. When you examine cross-tabulations in this book, always inspect the actual percentage differences to see how big a relationship really is. When you try to assess the strength of a relationship, also look at the size of Cramer's V. Any V greater than .30 should be considered a strong relationship, and any V between .10 and .30 should be considered a moderate relationship. A V under .10 should be considered a weak relationship. In the example above, V is .107, so we'd conclude that gender is moderately related to the importance placed on qualities for children to learn.

We can also get the table we just examined by using the AUTO-ANALYZER task. This task in effect combines the univariate and cross-tabulation procedures you've already seen in these pages. It first shows you the distribution of a *primary variable* you select and then allows you to choose one of nine demographic variables—sex, race, political party, marital relationship, religious preference, region, age, education, and income—to see what difference, if any, this demographic variable makes. It then gives

GROUPS AND ORGANIZATIONS

Tasks: Mapping, Scatterplot, Univariate, Historical Trends, Cross-tabulation
Data Files: NATIONS, STATES, GSS, HISTORY

Four centuries ago, the English poet John Donne wrote, "No man is an island." To avoid sexist language, today one might write, "No one is an island." However it's put, Donne's observation illustrates keen sociological insight. Except for the occasional hermit, all of us have ties to other people. We have so many ties, in fact, that we're members of many different types of groups. Group life and group membership are essential to social life and are key to appreciating the sociological perspective.

Two basic types of groups exist: primary groups and secondary groups. ***Primary groups*** are small and involve strong social bonds. Common examples include your family, groups of best friends, juvenile gangs, and perhaps sports teams and fraternities and sororities. Ideally, primary groups provide us with a sense of belonging and of our identity, emotional support, and practical needs. ***Secondary groups*** tend to be larger than primary groups and involve weaker and much more impersonal social bonds. Common examples include any classes you've taken, including the one for which you're reading this book, and any business in which you might have worked. Secondary groups are not nearly as important as primary ones for our emotional needs, but fulfill many of our practical needs and are certainly essential to any modern society. Many secondary groups are highly structured and are thus considered ***formal organizations***. Most formal organizations have clear lines of authority, explicit rules for behaving, and clearly defined roles for performing organizational tasks—in short, a division of labor.

This exercise explores some key aspects of group life and organizational membership around the world and within the United States. Its aim is to help you understand some of the key building blocks of modern society and thus to further your appreciation of the sociological perspective.

INTERNATIONAL EXAMPLES OF GROUP PERCEPTION AND MEMBERSHIP

Perhaps the most important primary group is the family, which we'll explore further in Exercise 10. However, it's also true that the family is more important to some people than to others within the same nation, and more important to the people of some nations than to those of other nations. Let's explore international differences in the importance attached to families. We'll use a measure, from the NATIONS data set, of the percent who say their families are "very important" in their lives.

> *Data File:* **NATIONS**
> *Task:* **Mapping**
> *Variable 1:* **128) FAMILY IMP**
> *View:* **Map**

FAMILY IMP -- PERCENT WHO SAY THE FAMILY IS VERY IMPORTANT IN THEIR LIVES (WVS)

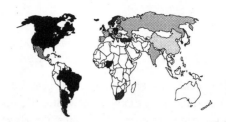

No clear pattern really emerges here, but it does seem that the nations of Eastern Europe and Asia are least likely to feel their family is very important.

Data File: **NATIONS**
Task: **Mapping**
Variable 1: **128) FAMILY IMP**
> *View:* **List: Rank**

RANK	CASE NAME	VALUE
1	Nigeria	94.00
2	South Korea	93.00
2	United States	93.00
4	Canada	92.00
5	Iceland	91.00
5	Brazil	91.00
5	Ireland	91.00
5	Argentina	91.00
5	Poland	91.00
10	South Africa	90.00

The residents of Nigeria, followed closely by those of South Korea, the United States, and Canada, are most likely to feel their family is very important; more than 90 percent in each nation report this view. At the other end, only 62 percent of the residents of China and Portugal feel this way. Note that even with these differences, the world's people are still quite apt to feel their families are very important.

What accounts for this international difference in the importance placed on families? Perhaps families are considered more important in nations that are more religious.

Data File: **NATIONS**
> *Task:* **Scatterplot**
> *Dependent Variable:* **128) FAMILY IMP**
> *Independent Variable:* **52) GOD IMPORT**
> *Display:* **Reg. Line**

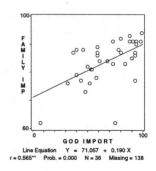

Line Equation   Y = 71.057 + 0.190 X
r = 0.565** Prob. = 0.000 N = 36 Missing = 138

Our hypothesis is supported: the more religious a nation's citizens, the more they believe the family is very important in their lives (r = .57**). Why should the more religious nations feel this way about the family?

The NATIONS data set includes a variable on membership in one type of voluntary association, a sports or recreation group. The percent belonging to such a group varies by nation.

      Data File: **NATIONS**
      ➤ *Task:* **Mapping**
      ➤ *Variable 1:* **125) DO SPORTS?**
      ➤ *View:* **Map**

DO SPORTS? -- PERCENT WHO SAY THEY BELONG TO A SPORTS OR RECREATION GROUP OR ORGANIZATION (WVS)

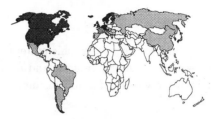

European nations have the highest rates of membership in sports or recreation groups.

What accounts for this international variation? Scholars say that education is a strong predictor of membership in many types of voluntary associations: the higher the education, the more associations joined. If this is true, then membership in sports or recreation groups should be higher in nations with more educated citizens.

      Data File: **NATIONS**
      ➤ *Task:* **Scatterplot**
➤ *Dependent Variable:* **125) DO SPORTS?**
➤ *Independent Variable:* **40) EDUCATION**
      ➤ *Display:* **Reg. Line**

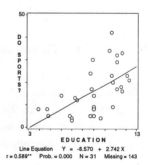

Line Equation   Y = -8.570 + 2.742 X
r = 0.589**   Prob. = 0.000   N = 31   Missing = 143

At the international level, membership in sports or recreation groups is linked to education (r = .59**).

ORGANIZATIONAL MEMBERSHIP IN THE UNITED STATES

One of the most controversial types of voluntary associations over the last decade or so has been the survivalist organization. An unknown number of people belong to these groups, which reportedly stock-pile weapons to defend themselves against perceived threats from the federal government. One rough measure of membership in, or at least interest in, survivalist organizations is the circulation of *Survive* magazine, a leading survivalist periodical. Let's examine its circulation per 1000 population.

➤ *Data File:* **STATES**
    ➤ *Task:* **Mapping**
➤ *Variable 1:* **112) SURVIVE**
    ➤ *View:* **Map**

SURVIVE -- 1983: CIRCULATION OF SURVIVE MAGAZINE (A LEADING SURVIVALIST PUBLICATION) PER 1000 POPULATION (ABC)

This is one of the clearest geographic patterns we've seen so far in the U.S. The circulation of *Survive* magazine—and presumably interest and membership in survivalist groups—is highest west of the Mississippi. What would you think accounts for this geographic pattern?

Data File: **STATES**
    Task: **Mapping**
Variable 1: **112) SURVIVE**
    ➤ *View:* **List: Rank**

RANK	CASE NAME	VALUE
1	Alaska	0.85
2	Nevada	0.60
3	Kansas	0.43
4	Colorado	0.38
5	New Mexico	0.37
5	Arizona	0.37
7	Hawaii	0.36
8	Oregon	0.32
9	Wyoming	0.30
10	Montana	0.26

Alaska leads the nation in *Survive* circulation, and New York and Missouri are tied with the lowest circulation.

A much different type of voluntary organization is the Junior League. This is a national organization with many local chapters and involves primarily conservative, upwardly mobile young women. We can examine the circulation of the *Junior Leaguer Review* magazine per 1000 population to get a rough idea of where Junior League membership is most common.

Data File: **STATES**
Task: **Mapping**
➤ Variable 1: **110) JR LEAGUE**
➤ View: **Map**

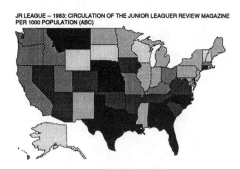

JR LEAGUE -- 1983: CIRCULATION OF THE JUNIOR LEAGUER REVIEW MAGAZINE PER 1000 POPULATION (ABC)

Generally, the circulation of the *Junior League Review* magazine (and presumably membership in the Junior League) is highest in the South and parts of the Midwest. Why do you think Junior League membership is so high in these parts of the country?

Georgia has the highest circulation rate, at 1.24 per 1000 population, while South Dakota has the lowest rate, at 0.01 per 1000 population.

These maps give us an idea of interest and membership in some very different kinds of organizations, but survey data are more often used to understand voluntary association and other secondary group membership in the United States. Accordingly, let's turn to the GSS, which contains several variables that will interest us.

A labor union certainly fits the definition of a secondary group. Let's see how many U.S. residents belong to a labor union.

➤ Data File: **GSS**
➤ Task: **Univariate**
➤ Primary Variable: **15) UNIONIZED?**
➤ View: **Pie**

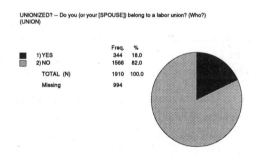

UNIONIZED? -- Do you (or your [SPOUSE]) belong to a labor union? (Who?) (UNION)

	Freq.	%
■ 1) YES	344	18.0
▨ 2) NO	1566	82.0
TOTAL (N)	1910	100.0
Missing	994	

Eighteen percent of Americans belong to a labor union.

We can use the HISTORICAL TRENDS task to see whether union membership has changed in the last quarter century.

> *Data File:* **HISTORY**
> > *Task:* **Historical Trends**
> > *Variable:* **1) UNIONIZED**

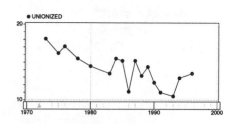

Percent of respondents belonging to a labor union

Open the HISTORY data file and select the HISTORICAL TRENDS task. Select 1) UNIONIZED as your trend variable.

Union membership has declined by a few percentage points since the early 1970s.

Whether or not we belong to labor unions, we all have opinions about them. The GSS asks how much confidence respondents have in organized labor.

> *Data File:* **GSS**
> > *Task:* **Univariate**
> *Primary Variable:* **24) LABOR?**
> > *View:* **Pie**

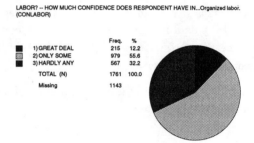

LABOR? -- HOW MUCH CONFIDENCE DOES RESPONDENT HAVE IN...Organized labor. (CONLABOR)

		Freq.	%
■	1) GREAT DEAL	215	12.2
▨	2) ONLY SOME	979	55.6
■	3) HARDLY ANY	567	32.2
	TOTAL (N)	1761	100.0
	Missing	1143	

About 12 percent of Americans say they have a "great deal" of confidence in organized labor, 56 percent "only some" confidence, and 32 percent "hardly any" confidence.

Has the percent expressing a great deal of confidence declined in the past 25 years along with union membership?

> *Data File:* **HISTORY**
> > *Task:* **Historical Trends**
> > *Trend 1:* **2) LABOR?**

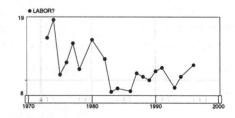

Percent expressing "great deal" of confidence in organized labor

Again, open the HISTORY data file and select the HISTORICAL TRENDS task. Select 2) LABOR? as your trend variable.

Discovering Sociology

Confidence in organized labor declined until the mid-1980s and then rose somewhat, but is still not back to its early-1970s level.

Many organizations benefit from people who volunteer their time to help out. Despite the importance of such "civic voluntarism" for these organizations, relatively few scholars have studied it. The GSS contains a series of questions asking respondents whether they've done volunteer work in various areas in the past 12 months. Let's look at responses to a few of these items to get an idea of how many Americans volunteer their time. We'll first see what percent have volunteered their time in a health-care organization, such as a hospital.

➤ *Data File:* **GSS**
➤ *Task:* **Univariate**
➤ *Primary Variable:* **152) VOL HEALTH**
➤ *View:* **Pie**

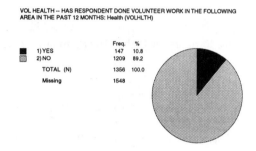

About 11 percent have done volunteer work in the health area.

Data File: **GSS**
Task: **Univariate**
➤ *Primary Variable:* **153) VOL EDUC**
➤ *View:* **Pie**

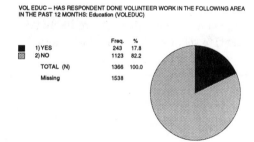

Almost 18 percent have volunteered in the education area.

What affects the likelihood of our volunteering in these areas? While we can't look here at every possible factor for each area, we can examine a few. As noted earlier, education is often an important factor in decisions to join voluntary organizations. Let's see whether more educated people are more likely to volunteer their time.

Data File: **GSS**

➤ *Task:* **Cross-tabulation**

➤ *Row Variable:* **152) VOL HEALTH**

➤ *Column Variable:* **6) EDUCATION**

➤ *View:* **Tables**

➤ *Display:* **Column %**

152) VOL HEALTH	6) EDUCATION			
	NO HS GRA	HS GRAD	SOME COL	COLL GRA
YES	4.6%	6.7%	13.8%	17.0%
NO	95.4%	93.3%	86.2%	83.0%
TOTAL	100.0%	100.0%	100.0%	100.0%

V=0.159**

College-educated people are more likely (17.0 percent) than those lacking a high school degree (4.6 percent) to volunteer for health organizations (V = .16**).

Data File: **GSS**

Task: **Cross-tabulation**

➤ *Row Variable:* **153) VOL EDUC**

➤ *Column Variable:* **6) EDUCATION**

➤ *View:* **Tables**

➤ *Display:* **Column %**

153) VOL EDUC	6) EDUCATION			
	NO HS GRA	HS GRAD	SOME COL	COLL GRA
YES	6.2%	11.9%	20.7%	29.2%
NO	93.8%	88.1%	79.3%	70.8%
TOTAL	100.0%	100.0%	100.0%	100.0%

V=0.220**

College-educated people are more likely (29.2 percent) than those without a high school degree (6.2 percent) to volunteer in the education area (V = .22**).

WORKSHEET

NAME:

COURSE:

DATE:

EXERCISE

4

REVIEW QUESTIONS

Based on the first part of this exercise, answer True or False to the following items:

Membership in sports and recreation groups is highest in European nations.	T F
Eastern European nations rank relatively low in the percent who say the family is very important in their lives.	T F
In the United States, survivalist interest seems higher east of the Mississippi than west of the Mississippi.	T F
Junior League membership seems higher in the Northeast than in other regions of the country.	T F
Education makes little difference in decisions to volunteer one's time for voluntary organizations.	T F
Confidence in organized labor is higher now than it was in the early 1970s.	T F

EXPLORIT QUESTIONS

1. Although the family as a primary group provides its members with emotional support and fills other important needs, it's also true that our families can be a source of great pain. The GSS asks whether respondents during the past month felt "really angry, irritated, or annoyed" about a situation that involved their families.

> ➤ *Data File:* **GSS**
> ➤ *Task:* **Univariate**
> ➤ *Primary Variable:* **97) ANGRY FAM**
> ➤ *View:* **Pie**

a. What percent felt angry, irritated, or annoyed in the past month about a family situation?

_____%

b. Is this figure lower than you expected, higher than you expected, or about what you expected? (Circle one.)

Lower

Higher

About what I expected

Exercise 4: Groups and Organizations

2. Let's see whether our gender affects whether we feel angry or annoyed about a family situation.

> Data File: **GSS**
> ➤ Task: **Cross-tabulation**
> ➤ Row Variable: **97) ANGRY FAM**
> ➤ Column Variable: **45) GENDER**
> ➤ View: **Tables**
> ➤ Display: **Column %**

a. What percent of women felt angry about a family situation? _____%

b. What percent of men felt angry about a family situation? _____%

c. Is V statistically significant?   Yes   No

d. Circle one of the following:

    1. Women are more likely than men to feel angry about a family situation.

    2. Men are more likely than women to feel angry about a family situation.

    3. Women and men are equally likely to feel angry about a family situation.

e. Looking at the statement you just circled, what sociological reason(s) might help explain what you found?

3. Should family income make a difference? What would you predict here?

> Data File: **GSS**
> Task: **Cross-tabulation**
> Row Variable: **97) ANGRY FAM**
> ➤ Column Variable: **7) FAM INCOME**
> ➤ View: **Tables**
> ➤ Display: **Column %**

a. What percent of the lowest income group felt angry about a family situation? _____%

b. What percent of the highest income group felt angry about a family situation? _____%

c. Is V statistically significant?   Yes   No

d. Circle one of the following:

1. Low-income people are more likely than high-income people to feel angry about a family situation.

2. High-income people are more likely than low-income people to feel angry about a family situation.

3. Income is not related to feeling angry about a family situation.

e. Were you surprised by what you found? Why or why not?

4. Is religiosity related to whether we feel angry or annoyed?

> Data File: **GSS**
> Task: **Cross-tabulation**
> Row Variable: **97) ANGRY FAM**
> ➤ Column Variable: **127) ATTEND**
> ➤ View: **Tables**
> ➤ Display: **Column %**

a. What percent of people who attend religious services weekly felt angry about a family situation?   _____%

b. What percent of people who never attend religious services felt angry about a family situation?   _____%

c. Is V statistically significant?   Yes   No

d. Circle one of the following:

1. People who attend religious services weekly are more likely than those who never attend to feel angry about a family situation.

2. People who never attend religious services are more likely than those who attend weekly to feel angry about a family situation.

3. Religious attendance is not related to feeling angry about a family situation.

e. Looking at the statement you just circled, what sociological reason(s) might help explain what you found?

5. Select another GSS variable that might be related to whether respondents feel angry about a family situation and obtain a cross-tabulation where 97) ANGRY FAM is the row variable and the variable you select is the column variable.

    a. What variable did you select?   _____

    b. What was the value of V?   V = _____

    c. Was V statistically significant?   Yes   No

    d. Summarize the relationship that you found in your cross-tabulation.

6. Based on the results of these worksheet examples, what social backgrounds make it more likely that we'll feel angry, irritated, or annoyed about a family situation?

7. How do these results overall illustrate the sociological perspective?

8. Earlier we found that international variation in belonging to sports and recreation groups is related to the educational level of a nation's population. Perhaps wealthier nations are also more likely to have such groups and to have people who can afford to belong to them. If so, wealthier nations should have higher rates of membership in these groups than poorer nations.

    > *Data File:* **NATIONS**
    > *Task:* **Scatterplot**
    > *Dependent Variable:* **125) DO SPORTS?**
    > *Independent Variable:* **29) $ PER CAP**
    > *Display:* **Reg. Line**

    a. Taking into account the strength and statistical significance of r, what conclusion would you draw from this scatterplot?

b. How might the results of this scatterplot help you understand the relationship, seen earlier, between educational level and membership in sports and recreation groups?

9. One type of formal organization is the coercive organization, which, as the name implies, includes people who are forced to "belong to" the organization. Your STATES data set lists the percent of each state's population living in one type of coercive organization, the prison.

> ➤ *Data File:* **STATES**
> ➤ *Task:* **Mapping**
> ➤ *Variable 1:* **50) %IN PRISON**
> ➤ *View:* **Map**

a. Which two regions generally have the highest rates of incarceration (being behind bars)? (Circle one.)

                                                 Northeast and Midwest

                                                      South and West

                                                South and Midwest

b. The regional imprisonment rates you see might be explained by one or more of the following: (1) regional differences in crime; (2) regional differences in the chances of going to prison once you've been arrested; (3) racial bias in arrest and punishment. Which one of these three reasons do you think accounts most for the regional imprisonment rates you found? Why?

c. Suppose you have an adequate measure of regional differences in crime and want to determine whether such differences account for the regional differences in imprisonment. You obtain a scatterplot demonstrating the relationship between these two variables. If regional variation in crime does account for regional variation in imprisonment, what should the scatterplot look like? Explain your answer.

10. Earlier we looked at membership in organized labor. Historically, organized labor has found it difficult to gain a foothold in the South. Our hypothesis is that Southerners are less likely than people in other regions to belong to unions.

> ➤ Data File: **GSS**
> ➤ Task: **Cross-tabulation**
> ➤ Row Variable: **15) UNIONIZED**
> ➤ Column Variable: **62) REGION**
> ➤ View: **Tables**
> ➤ Display: **Column %**

    a. Which region in the table has the lowest membership in labor unions? (Circle one.)

Northeast

South

Midwest

West

    b. Taking V and statistical significance into account, are the people in the region you indicated above less likely than people in other regions to belong to labor unions?   Yes   No

11. Based on the results of the preceding example, do you think Southerners should have less confidence in organized labor than people in other regions? Let's see.

> Data File: **GSS**
> Task: **Cross-tabulation**
> ➤ Row Variable: **24) LABOR?**
> ➤ Column Variable: **62) REGION**
> ➤ View: **Tables**
> ➤ Display: **Column %**

    a. Taking V into account, do Southerners have less confidence in organized labor than people in other regions? (Circle one.)   Yes   No

    b. Did the results of this table surprise you? Why or why not?

◆ EXERCISE 5 ◆

DEVIANCE, CRIME, AND SOCIAL CONTROL

Tasks: Mapping, Scatterplot, Univariate, Historical Trends, Cross-tabulation
Data Files: NATIONS, STATES, GSS, HISTORY

If all societies have norms, or standards guiding behavior, it's also true that all societies have deviance, or violations of these norms. Emile Durkheim wrote long ago that a society without deviance is impossible, because there will always be some people who sometimes violate norms. The collective conscience, said Durkheim, is never strong enough to prevent all norm violation. If it were that strong, he added, the society would be very stagnant, because this would also be a society where people were unable to think creatively. Creative thinking presupposes a society where the collective conscience is weak enough that people are able to be independent thinkers. Yet such a society will also have people who choose to violate norms. You cannot have the first type of society without also having the second.

Deviance takes many forms. Some involves violations of norms that, in the grand scheme of things, are not that serious, for example dyeing your hair green or jaywalking, whereas other kinds of deviance involve much more important norms, such as murder, rape, and robbery. The latter types typically involve violations of formal, written norms, or criminal laws, and thus are more commonly called crimes.

This brief description might imply that something automatically is or is not deviant, but sociologists emphasize that what is considered deviant often depends on the reactions of others and the circumstances in which any particular act occurs. In this context, consider killing. If you kill someone because you're angry at him or her, you've committed a homicide, perhaps our most serious "street crime." But if you kill someone on the battlefield, you're instead doing your duty for your country, and if you kill many people on the battlefield, you may be considered a hero. In each case, killing has occurred; in the first case you go to prison or may even be executed, whereas in the second case you may get a medal.

This chapter examines some of the correlates of crime and deviance in the United States and around the world. It also explores the correlates of some attitudes toward these behaviors and the punishment of offenders who commit them. The chapter will help us once again to see what influence, if any, social backgrounds have on behaviors and attitudes.

A CROSS-CULTURAL LOOK AT DEVIANCE

As we look around the globe, we see much variation in approval and disapproval of various kinds of behaviors often considered deviant. Our NATIONS data set contains measures of views on many of these behaviors. Let's look at international opinion on suicide by using a measure of the percent who say suicide is never acceptable.

> *Data File:* **NATIONS**
> *Task:* **Mapping**
> *Variable 1:* **115) SUICIDE NO**
> *View:* **Map**

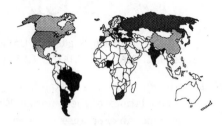

SUICIDE NO -- PERCENT WHO THINK SUICIDE IS NEVER OK (WVS)

The darker the color, the greater the disapproval of suicide. Notice that disapproval appears highest in the Southern Hemisphere and some parts of Europe and lowest in other parts of Europe.

Data File:	**NATIONS**	
Task:	**Mapping**	
Variable 1:	**115) SUICIDE NO**	
> *View:*	**List: Rank**	

RANK	CASE NAME	VALUE
1	Brazil	89.00
2	Argentina	82.00
3	India	81.00
3	Turkey	81.00
3	Chile	81.00
6	Romania	76.00
7	Poland	74.00
8	Nigeria	73.00
9	Portugal	72.00
10	South Africa	71.00

Disapproval is highest in Brazil, where 89 percent think suicide is never acceptable. In contrast, only 31 percent of people in the Netherlands feel this way. In the United States, 61 percent think suicide is never acceptable.

What accounts for these international differences? One possible answer might lie in education: perhaps nations with lower levels of education are more likely to hold traditional attitudes toward suicide, and nations with higher levels of education less likely. If so, we would expect a negative correlation between education and disapproval of suicide.

> *Data File:* **NATIONS**
> *Task:* **Scatterplot**
> *Dependent Variable:* **115) SUICIDE NO**
> *Independent Variable:* **40) EDUCATION**
> *Display:* **Reg. Line**

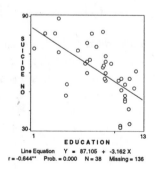

Line Equation $Y = 87.105 + -3.162 X$
$r = -0.644**$ Prob. = 0.000   N = 38   Missing = 136

As predicted, the higher the education level of a nation, the lower its disapproval of suicide (r = −.64**).

Let's turn now from international opinions on suicide to international differences in actual suicide rates.

> *Data File:* **NATIONS**
> ➤ *Task:* **Mapping**
> ➤ *Variable 1:* **114) SUICIDE**
> ➤ *View:* **Map**

SUICIDE -- SUICIDES PER 100,000 (MOST RECENT YEAR) (IP)

Suicide rates tend to be highest in Europe.

> *Data File:* **NATIONS**
> *Task:* **Mapping**
> *Variable 1:* **114) SUICIDE**
> ➤ *View:* **List: Rank**

RANK	CASE NAME	VALUE
1	Hungary	42.6
2	Sri Lanka	30.0
3	Finland	29.9
4	Denmark	26.6
5	Austria	23.1
6	Belgium	22.3
7	Switzerland	21.0
8	Russia	20.8
9	France	19.8
10	Luxemburg	18.6

Hungary leads the world with a suicide rate of 42.6 per 100,000 population. At 0.2 per 100,000, Iran and Syria have the lowest rate. The U.S. rate is 11.7.

Should more religious nations have lower suicide rates than less religious nations?

> *Data File:* **NATIONS**
> ➤ *Task:* **Scatterplot**
> ➤ *Dependent Variable:* **114) SUICIDE**
> ➤ *Independent Variable:* **52) GOD IMPORT**
> ➤ *Display:* **Reg. Line**

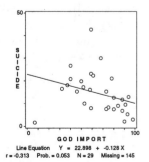

Line Equation   Y = 22.898 + -0.128 X
r = -0.313   Prob. = 0.053   N = 29   Missing = 145

Although this correlation is not statistically significant, notice that in the lower left-hand corner of the scatterplot is a dot that is far away from the other dots because its religious attendance and suicide levels are both much lower than those for other nations. This dot is called an *outlier*. Because its presence in the scatterplot may be distorting the correlation coefficient, we should consider removing it from the plot. To do so, click the [Outlier] option. When you do so, you'll see this dot highlighted, and information will appear identifying it as China. This information also indicates that the new correlation coefficient after removing China would be −.54**, a significantly higher correlation than the one we started with. Go ahead and click the [Remove] button. With the outlier of China removed, we now see that religiosity is rather strongly related to suicide rates across the world.

CRIME AND DEVIANCE IN THE UNITED STATES

One of the most important questions in the United States is why people commit crime and deviance. Every year the FBI reports the amount and rate of various types of crime, most notably homicide, rape, aggravated assault, robbery, burglary, larceny, and motor vehicle theft. This report is called the Uniform Crime Reports (UCR). By convention, the number of crimes in a state is divided by the state's population and then multiplied by 100,000 to yield a crime rate per 100,000 population. This allows us to compare crime rates among states with very different populations. Let's examine some state-level correlates of violent crime (homicide, rape, aggravated assault, and robbery) in the United States to get a sociological sense of why people commit such crimes. We'll first map violent crime.

➤ *Data File:* **STATES**
➤ *Task:* **Mapping**
➤ *Variable 1:* **95) V.CRIME**
➤ *View:* **Map**

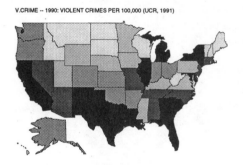

V.CRIME -- 1990: VIOLENT CRIMES PER 100,000 (UCR, 1991)

Notice that violent crime rates are not randomly distributed throughout the U.S. Instead, they are patterned geographically: the rates are highest in the South, the Far West, and a few northeastern states, and lowest in the upper Midwest and northern New England. Why does this pattern exist?

One possible explanation lies in whether a state is urban or rural. Urban areas obviously have a lot of people living very close together. These crowded conditions can prompt violent crime because they offer a ready supply of potential victims for would-be criminals and may also cause tempers to flare. Let's see whether the more urban states have higher violent crime rates.

Data File: **STATES**
➤ Task: **Scatterplot**
➤ Dependent Variable: **95) V.CRIME**
➤ Independent Variable: **11) %URBAN**
➤ Display: **Reg. Line**

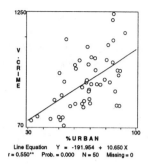

Line Equation   Y = -191.954 + 10.650 X
r = 0.550**   Prob. = 0.000   N = 50   Missing = 0

Our hypothesis is supported: the more urban a state, the higher its crime rate (r = .55**).

Another possible explanation comes from ***social disorganization*** theory, which attributes high community crime rates to the weakening of social institutions such as religion and family by severe poverty, overcrowding, and other problems. One factor often cited here is the single-parent household. Areas with high levels of such households are apt to have higher crime rates for several reasons. First, they have fewer networks of "informal" social control and thus are less able than other areas to supervise adolescents and to be on the lookout for strangers. Second, areas with many single-parent households are more vulnerable to criminal victimization because single adults are more likely to be alone when outside, whether going to work or engaging in leisure-time activity, and are thus more vulnerable to personal crimes such as robbery, rape, and murder. Third, and this is the subject of much debate, single-parent households may do a poorer job of raising their children, who are thus more likely to grow into criminals.

With these reasons in mind, let's see whether the states with the highest violent crime rates also tend to be the states with the highest levels of single-parent households. Since most such households are headed by women, we'll use a measure of the percent of households that are female-headed with children present.

Data File: **STATES**
Task: **Scatterplot**
Dependent Variable: **95) V.CRIME**
➤ Independent Variable: **40) F HEAD W/C**
➤ Display: **Reg. Line**

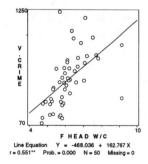

Line Equation   Y = -468.036 + 162.767 X
r = 0.551**   Prob. = 0.000   N = 50   Missing = 0

States with higher rates of single-parent households have higher violent crime rates (r = .55**).

Now let's switch to the GSS data set to consider U.S. public opinion on crime and deviance issues. We'll start with a GSS question as to whether marijuana use should be made legal.

> ➤ *Data File:* **GSS**
> ➤ *Task:* **Univariate**
> ➤ *Primary Variable:* **119) GRASS?**
> ➤ *View:* **Pie**

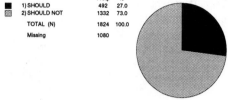

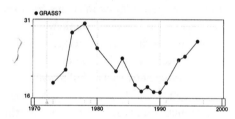

Twenty-seven percent of the GSS respondents think marijuana use should be legalized.

> ➤ *Data File:* **HISTORY**
> ➤ *Task:* **Historical Trends**
> ➤ *Variable:* **3) GRASS?**

Percent saying marijuana should be made legal

This percent has certainly changed during the past quarter century. It rose through the 1970s, declined steadily until the late 1980s, then rose again during the 1990s.

Earlier we saw that the more educated nations were more likely to accept suicide. Let's see whether the more educated Americans are more likely to think marijuana use should be legalized. Do you think we'll find support for marijuana legalization rising with education?

> ➤ *Data File:* **GSS**
> ➤ *Task:* **Cross-tabulation**
> ➤ *Row Variable:* **119) GRASS?**
> ➤ *Column Variable:* **6) EDUCATION**
> ➤ *View:* **Tables**
> ➤ *Display:* **Column %**

6) EDUCATION

119) GRASS?	NO HS GRA	HS GRAD	SOME COL	COLL GRA
SHOULD	23.6%	25.5%	27.5%	30.3%
SHOULD NOT	76.4%	74.5%	72.5%	69.7%
TOTAL	100.0%	100.0%	100.0%	100.0%

V=0.052

Taking into account statistical significance, education is not related to support for marijuana legalization (V = .05). We cannot assume that support for marijuana legalization rises with education.

What about religiosity? Do you think we'll find that less religious people are more likely than more religious people to favor the legalization of marijuana?

Discovering Sociology

<table>
<tr><td></td><td></td><td colspan="3">127) ATTEND</td></tr>
<tr><td></td><td></td><td>NEVER</td><td>MONTH/YR</td><td>WEEKLY</td></tr>
</table>

Data File:	**GSS**		
Task:	**Cross-tabulation**		
Row Variable:	**119) GRASS?**		
➤ Column Variable:	**127) ATTEND**		
➤ View:	**Tables**		
➤ Display:	**Column %**		

119) GRASS?	NEVER	MONTH/YR	WEEKLY
SHOULD	41.2%	30.0%	14.3%
SHOULD NOT	58.8%	70.0%	85.7%
TOTAL	100.0%	100.0%	100.0%

V=0.208**

People who never attend religious services are more likely (41.2 percent) than those who attend weekly (14.3 percent) to think that marijuana use should be legalized (V = .21**).

One reason crime gets so much attention these days is that so many people are concerned about it. The GSS asks, "Is there any area right around here—that is, within a mile—where you would be afraid to walk alone at night?" This is a common measure of fear of crime in the criminological literature. Let's see how many Americans are afraid of crime.

Data File:	**GSS**
➤ Task:	**Univariate**
➤ Primary Variable:	**121) FEAR WALK**
➤ View:	**Pie**

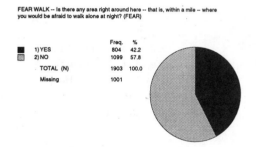

FEAR WALK -- Is there any area right around here -- that is, within a mile -- where you would be afraid to walk alone at night? (FEAR)

		Freq.	%
■	1) YES	804	42.2
▨	2) NO	1099	57.8
	TOTAL (N)	1903	100.0
	Missing	1001	

About 42 percent say they'd be afraid to walk near their homes at night.

Has this percent changed during the past 25 years?

➤ Data File:	**HISTORY**
➤ Task:	**Historical Trends**
➤ Variable:	**4) FEAR WALK**

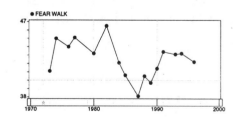

Percent saying they'd be afraid to walk near home at night

Fear of crime declined during the 1980s but is back to about its 1970s level.

Many aspects of our social backgrounds might affect the degree to which we're afraid of crime. Let's look first at gender. Do you think women will be more afraid than men or men more afraid than women, or do you think no gender difference exists?

> ➤ *Data File:* **GSS**
> ➤ *Task:* **Cross-tabulation**
> ➤ *Row Variable:* **121) FEAR WALK**
> ➤ *Column Variable:* **45) GENDER**
> ➤ *View:* **Tables**
> ➤ *Display:* **Column %**

121) FEAR WALK	45) GENDER FEMALE	MALE
YES	55.4%	25.8%
NO	44.6%	74.2%
TOTAL	100.0%	100.0%

V=0.298**

Women are more likely (55.4 percent) than men (25.8 percent) to say they'd be afraid to walk alone at night in their neighborhood. The relationship is very strong (V = .30**).

Now let's look at race. Most scholars agree that African Americans have higher rates than whites of crime and victimization. We thus hypothesize that they should be more likely than whites to fear crime in their neighborhood.

> *Data File:* **GSS**
> *Task:* **Cross-tabulation**
> *Row Variable:* **121) FEAR WALK**
> ➤ *Column Variable:* **42) WHTE/AFRAM**
> ➤ *View:* **Tables**
> ➤ *Display:* **Column %**

121) FEAR WALK	42) WHTE/AFRAM WHITE	AFRICANA
YES	40.1%	55.8%
NO	59.9%	44.2%
TOTAL	100.0%	100.0%

V=0.114**

African Americans are more likely (55.8 percent) than whites (40.1 percent) to be afraid to walk alone at night (V = .11**).

Since women are so much more afraid of crime than men, let's hypothesize that they should be more likely than men to favor additional government spending on law enforcement.

> *Data File:* **GSS**
> *Task:* **Cross-tabulation**
> *Row Variable:* **125) SP.POLICE**
> ➤ *Column Variable:* **45) GENDER**
> ➤ *View:* **Tables**
> ➤ *Display:* **Column %**

125) SP.POLICE	45) GENDER FEMALE	MALE
MORE	58.2%	57.8%
SAME	35.7%	34.1%
LESS	6.1%	8.0%
TOTAL	100.0%	100.0%

V=0.038

No gender difference at all (V = .04)! How would you explain this surprising result?

Since African Americans are more afraid of crime than whites, should we hypothesize that they should also be more likely than whites to favor additional spending on law enforcement? Or does our surprising result for gender give you pause? Only one way to find out:

Data File: **GSS**
Task: **Cross-tabulation**
Row Variable: **125) SP.POLICE**
➤ Column Variable: **42) WHTE/AFRAM**
➤ View: **Tables**
➤ Display: **Column %**

42) WHTE/AFRAM		
	WHITE	AFRICANA
125) MORE	58.7%	55.0%
SP SAME	35.3%	33.7%
POLICE LESS	6.0%	11.2%
TOTAL	100.0%	100.0%

V=0.073*

Despite their lower fear of crime, whites turn out to be a bit more likely than African Americans to favor more government spending on law enforcement (V = .07*). How would you explain this result?

WORKSHEET

NAME:

COURSE:

DATE:

EXERCISE

5

REVIEW QUESTIONS

Based on the first part of this exercise, answer True or False to the following items:

At the international level, the higher the education, the lower the disapproval of suicide.	T	F
At the international level, the higher the religiosity, the lower the suicide rate.	T	F
In the United States, violent crime is higher in the Northeast than in any other region of the country.	T	F
In the United States, more-urban states have higher crime rates than more-rural states.	T	F
In the GSS, support for the legalization of marijuana has risen steadily since the early 1970s.	T	F

EXPLORIT QUESTIONS

1. In the area of deviance and crime, gun control and the death penalty are two of the most controversial issues in the United States today. Let's start with gun control.

> ➤ Data File: **GSS**
> ➤ Task: **Cross-tabulation**
> ➤ Row Variable: **118) GUN LAW?**
> ➤ Column Variable: **45) GENDER**
> ➤ View: **Tables**
> ➤ Display: **Column %**

a. Who is more likely to favor gun control?   Women   Men

b. What would be a sociological explanation for the gender difference you found?

2.   Does race affect support for gun control legislation?

> *Data File:*   **GSS**
> *Task:*   **Cross-tabulation**
> *Row Variable:*   **118) GUN LAW?**
> ➤ *Column Variable:*   **42) WHTE/AFRAM**
> ➤ *View:*   **Tables**
> ➤ *Display:*   **Column %**

a.   Taking into account statistical significance, are there racial differences in the United States in support for gun control?   Yes   No

b.   Did this result surprise you? Why or why not?

3.   What about income? Be sure to check for statistical significance.

> *Data File:*   **GSS**
> *Task:*   **Cross-tabulation**
> *Row Variable:*   **118) GUN LAW?**
> ➤ *Column Variable:*   **7) FAM INCOME**
> ➤ *View:*   **Tables**
> ➤ *Display:*   **Column %**

a.   The higher the income, the greater the support for gun control.   T   F

b.   Less than half of low-income families support gun control.   T   F

c.   More than 80 percent of medium- and high-income families favor gun control.   T   F

4.   We turn now to the death penalty.

> *Data File:*   **GSS**
> *Task:*   **Cross-tabulation**
> ➤ *Row Variable:*   **117) EXECUTE?**
> ➤ *Column Variable:*   **45) GENDER**
> ➤ *View:*   **Tables**
> ➤ *Display:*   **Column %**

a.   Is there a gender difference in support for the death penalty?   Yes   No

b. If so (and it might not be so), who favors the death penalty more?

Women

Men

Neither

c. What would be a sociological explanation for the result you find in the table?

5. Should race affect support for the death penalty?

> Data File: **GSS**
> Task: **Cross-tabulation**
> Row Variable: **117) EXECUTE?**
> ➤ Column Variable: **42) WHITE/AFRAM**
> ➤ View: **Tables**
> ➤ Display: **Column %**

a. African Americans are more likely than whites to support the death penalty.   T   F

b. More than half of all African Americans support the death penalty.   T   F

c. About 19 percent of whites oppose the death penalty.   T   F

d. What would be a sociological explanation for the result you find in the table?

6. Now let's go back to the NATIONS data set.

> ➤ Data File: **NATIONS**
> ➤ Task: **Mapping**
> ➤ Variable 1: **112) PROSTITUTE**
> ➤ View: **List: Rank**

a. Which country has the highest percent who think prostitution is never OK?   _____

b. Which country has the lowest percent?   _____

c. What is the percent for the United States? _____%

d. Fill in the percentages for the following nations:

Chile _____%
Sweden _____%
Canada _____%
Mexico _____%

7. Should disapproval of prostitution be higher in the more religious nations?

Data File: **NATIONS**
➤ Task: **Scatterplot**
➤ Dependent Variable: **112) PROSTITUTE**
➤ Independent Variable: **53) PRAY?**
➤ Display: **Reg. Line**

a. What is the value of r? r = _____

b. Is r statistically significant? Yes   No

c. If we hypothesize that the more religious nations should be more likely to think that prostitution is never OK, do the data support the hypothesis? Yes   No

d. What other variable about a nation do you think should affect the degree to which its citizens oppose prostitution? Why?

8. Many scholars think that social support networks, including friendships, can help individuals deal with personal problems and perhaps even reduce their likelihood of committing suicide. The NATIONS data set has a measure of each nation's population who say friends are very important in their lives. If friendships do matter, it stands to reason that nations reporting lower levels of friends' importance should have higher suicide rates. If so, we'd expect to see a negative correlation between this variable and suicide.

Data File: **NATIONS**
Task: **Scatterplot**
➤ Dependent Variable: **114) SUICIDE**
➤ Independent Variable: **66) FRIENDS?**
➤ Display: **Reg. Line**

a. Summarize the results of this scatterplot.

b. Do the results of this scatterplot support our hypothesis?   Yes   No

9. In the preliminary section we saw a religiosity difference in the GSS in support for the legalization of marijuana, with less-religious people more in favor of legalization than more-religious people. Will gender make a difference? Since males are socialized to be more daring and assertive, let's hypothesize that they'll be more likely than women to favor marijuana legalization.

> ➤ *Data File:*   **GSS**
> ➤ *Task:*   **Cross-tabulation**
> ➤ *Row Variable:*   **119) GRASS**
> ➤ *Column Variable:*   **45) GENDER**
> ➤ *View:*   **Tables**
> ➤ *Display:*   **Column %**

a. What percent of men favor legalization?   _____

b. What percent of women?   _____

c. Taking V into account, is the hypothesis supported?   Yes   No

10. Continuing with marijuana legislation, what difference should urban vs. rural residence make? Much research finds that urban residents are more tolerant of unconventional behaviors. We'll thus hypothesize that they should be more likely than rural residents to favor the legalization of marijuana. To test this hypothesis, we'll use a GSS measure that indicates whether respondents have lived in an urban or rural area all their lives; respondents who have lived in both areas are excluded from the analysis.

> *Data File:*   **GSS**
> *Task:*   **Cross-tabulation**
> *Row Variable:*   **119) GRASS**
> ➤ *Column Variable:*   **94) URBAN?**
> ➤ *View:*   **Tables**
> ➤ *Display:*   **Column %**

a. Complete the following sentence: Urban residents are more likely (_____ percent) than rural
    residents (_____ percent) to favor the legalization of marijuana.

b. What is it about living in an urban area that should lead to a more tolerant
    view of marijuana use?

11. In the preliminary section of this exercise, we saw that women are more afraid than men of walking
    near their homes at night, and African Americans are more afraid than whites. If you combine these
    two sets of findings, black women should be more afraid than black men, white women, or white
    men. Let's find out.

<blockquote>

Data File:	**GSS**
Task:	**Cross-tabulation**
➤ Row Variable:	**121) FEAR WALK**
➤ Column Variable:	**44) RACEGENDER**
➤ View:	**Tables**
➤ Display:	**Column %**

</blockquote>

a. List the percent for each race/gender combination that answers "yes" to the question of fear of
    walking around home at night.

Black women	_____%
White women	_____%
Black men	_____%
White men	_____%

b. Looking at V, is the hypothesis supported?   Yes   No

c. This table indicates that race and gender interact to produce fear of crime. In the space below,
    write a brief essay explaining how and why the table indicates this interaction effect. (Hint: First
    see what difference race makes when you look only at women and then men, and then see what
    difference gender makes when you look only at whites and then blacks.)

SOCIAL STRATIFICATION

Tasks: Mapping, Scatterplot, Cross-tabulation
Data Files: NATIONS, STATES, GSS

In today's world, most societies are ***stratified***, meaning that they're characterized by an unequal distribution of resources that the society values. This unequal distribution yields a social hierarchy, with some people ranked at the top of the hierarchy, others ranked at the bottom, and the rest somewhere in between. As Max Weber pointed out, the resources in the modern world that most determine social ranking are wealth, power, and prestige. At the top of stratification systems today, then, are the people who have the most wealth, power, and/or prestige, while at the bottom are the people who have the least. Although the people with the most wealth usually also have much power and prestige and vice versa, this is not always true, as Weber also pointed out. For example, a big-time drug dealer might be very wealthy, but obviously has little prestige outside of the criminal world. Professional athletes are also quite wealthy, but again lack power in the political sense that Weber meant it.

Social stratification has important consequences for ***life chances***, or the outcomes we can expect in life's endeavors. Some common negative outcomes include poor physical and mental health, low educational achievement, and divorce and other family problems. In all these areas, according to much research, the poor fare far worse than the rich.

If some individuals are much wealthier than other individuals, it is also true that some nations are much wealthier than other nations. We call this difference in wealth ***global stratification***. The nations at the bottom of this social hierarchy are desperately poor, with hunger and disease rampant. Many of the differences in life chances between the poor and the non-poor in individual societies also exist between poor and rich societies at the global level.

This chapter examines some key features of stratification throughout the world and within the United States. Although our primary focus will be the correlates, including several life chances, of both types of stratification, we will also look at views on such issues as government spending and attitudes toward the poor.

GLOBAL STRATIFICATION

Let's begin our look at global stratification by first getting a picture of where the richest and poorest nations lie, using a common measure of the annual national product per capita.

> *Data File:* **NATIONS**
> *Task:* **Mapping**
> *Variable 1:* **29) $ PER CAP**
> *View:* **Map**

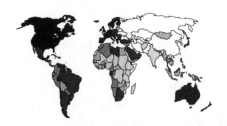

$ PER CAP -- ANNUAL NATIONAL PRODUCT PER CAPITA (TWF 1994)

The darker the color, the higher the nation's annual national product per capita. The nations in North America and Western Europe tend to be the wealthiest on this measure, and nations in Africa and Asia the poorest. The nations of the world are, indeed, stratified.

A nation's wealth or poverty has important implications for its people's life chances in almost every realm of human existence. To illustrate this, let's examine infant mortality, usually measured as the number of infant deaths before age 1 per 1000 births.

Data File: **NATIONS**
Task: **Mapping**
> *Variable 1:* **11) INF. MORTL**
> *View:* **Map**

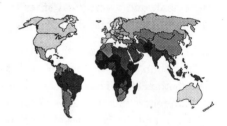

INF. MORTL -- NUMBER OF INFANT DEATHS PER 1,000 BIRTHS (TWF 1994)

Infant mortality is highest in Africa and parts of Asia.

Data File: **NATIONS**
Task: **Mapping**
Variable 1: **11) INF. MORTL**
> *View:* **List: Rank**

RANK	CASE NAME	VALUE
1	Somalia	162.70
2	Afghanistan	158.90
3	Western Sahara	155.50
4	Angola	148.60
5	Sierra Leone	145.00
6	Malawi	141.90
7	Guinea	141.70
8	Central African Republic	138.70
9	Chad	134.00
10	Mozambique	131.40

In the nations with the highest infant mortality rates, more than 140, or 14 percent, of every 1000 infants die before age 1. In the nations with the lowest rates, fewer than 6, or under 1 percent, of every 1000 infants die before age 1.

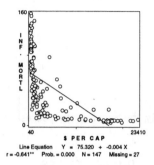

Data File: **NATIONS**
➤ Task: **Scatterplot**
➤ Dependent Variable: **11) INF. MORTL**
➤ Independent Variable: **29) $ PER CAP**
➤ Display: **Reg. Line**

The lower the national wealth, the higher the infant mortality. A baby born in a very poor nation has a much greater chance of dying young.

It's not a matter of life and death, but people in the United States take telephones, televisions, and cars for granted. These American "necessities of life" are, however, virtually unknown in many parts of the world. Let's use a scatterplot to show the strong association between national wealth and the presence of one of these items.

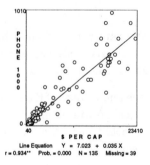

Data File: **NATIONS**
Task: **Scatterplot**
➤ Dependent Variable: **37) PHONE 1000**
➤ Independent Variable: **29) $ PER CAP**
➤ Display: **Reg. Line**

Wealthier nations have many more phones than poorer nations. The correlation, r, is an extremely high .93**.

As this scatterplot indicates, what is considered a "necessity of life" in the United States and other wealthy nations is an unaffordable luxury in much of the rest of the world.

STRATIFICATION IN THE UNITED STATES

To begin our look at U.S. stratification, let's first see which states are the poorest. We'll use the percent of the state's population below the poverty level.

➤ *Data File:* **STATES**
➤ *Task:* **Mapping**
➤ *Variable 1:* **55) %POOR**
➤ *View:* **Map**

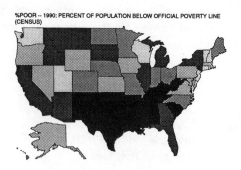

%POOR -- 1990: PERCENT OF POPULATION BELOW OFFICIAL POVERTY LINE (CENSUS)

Generally, the South is the poorest region of the country.

Data File: **STATES**
Task: **Mapping**
Variable 1: **55) %POOR**
➤ *View:* **List: Rank**

RANK	CASE NAME	VALUE
1	Mississippi	25.0
2	Louisiana	23.2
3	New Mexico	21.1
4	Arkansas	19.8
5	Alabama	19.1
6	Tennessee	17.8
7	West Virginia	17.2
8	Texas	17.0
8	Kentucky	17.0
10	South Carolina	16.2

The poverty rate ranges from a high of 25.0 percent in Mississippi to a low of 4.3 percent in Connecticut. Mississippi's rate is a full six times greater than Connecticut's. Where does your home state rank?

What are the implications of the states' poverty levels for the life chances of their residents? Earlier we saw that global stratification is linked to international differences in infant mortality. Does a similar relationship hold true within the United States? Let's first see which states rank highest and lowest on this life chance.

Data File: **STATES**
Task: **Mapping**
➤ *Variable 1:* **44) CHLD MORTL**
➤ *View:* **Map**

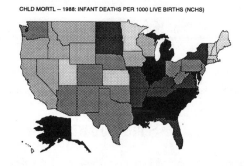

CHLD MORTL -- 1988: INFANT DEATHS PER 1000 LIVE BIRTHS (NCHS)

The Southeast and a few other states have the highest infant mortality rates.

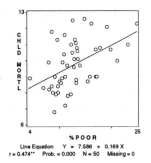

Data File: **STATES**
➤ Task: **Scatterplot**
➤ Dependent Variable: **44) CHLD MORTL**
➤ Independent Variable: **55) %POOR**
➤ Display: **Reg. Line**

The poorer the state, the higher its infant mortality (r = .47**). Although the United States has a lower infant mortality rate than most other nations, poverty at the state level is still linked to the chances of an infant's dying before age 1.

We now turn to the General Social Survey to explore the roots and impact of U.S. stratification at the individual level. Since its inception, the United States has been thought to be a land of equal opportunity for all, a country where people who work hard enough can pull themselves up by their bootstraps, as the saying goes, and achieve the American dream. Compared to most countries in the world, the United States does enjoy more social mobility. Still, sociologists emphasize that the chances of achieving the American dream depend to a large degree on where one starts out in life.

To illustrate this point, let's examine the relationship between our family's financial circumstances when we were youths and our own family's income when we're adults. The independent variable, 19) PARS.PRESG, from the GSS indicates whether both of the respondent's parents were in low-prestige jobs (as determined by a national ranking of prestige scores assigned to a range of jobs), medium-prestige jobs, or high-prestige jobs. These categories roughly correspond to low-paying, medium-paying, and high-paying jobs, respectively.

➤ Data File: **GSS**
➤ Task: **Cross-tabulation**
➤ Row Variable: **7) FAM INCOME**
➤ Column Variable: **19) PARS.PRESG**
➤ View: **Tables**
➤ Display: **Column %**

19) PARS.PRESG

7) FAM INCOME	BOTH LOW	BOTH MEDI	BOTH HIGH
$1K-$19999	33.5%	25.2%	13.9%
$20K-39999	39.1%	32.6%	33.5%
$40K & UP	27.4%	42.2%	52.6%
TOTAL	100.0%	100.0%	100.0%

V=0.174**

Our family's financial state when we were young does predict the income our families enjoy when we're adults (V = .17**). Where you start out in life apparently does influence where you end up. Note that the table does not tell us *why* this is so.

The GSS includes several variables measuring different aspects of how well the respondents think their lives are going. Let's examine the relationship between family income and two of these variables.

Exercise 6: Social Stratification

81

We'll start with a variable asking respondents, "Taken all together, how would you say things are these days—would you say that you are very happy, pretty happy, or not too happy?"

Data File: **GSS**
Task: **Cross-tabulation**
➤ Row Variable: **132) HAPPY?**
➤ Column Variable: **7) FAM INCOME**
➤ View: **Tables**
➤ Display: **Column %**

7) FAM INCOME			
	$1K-$19999	$20K-39999	$40K & UP
132) VERY HAPPY	22.4%	29.7%	35.7%
PRET.HAPPY	57.3%	59.6%	57.4%
NOT TOO	20.3%	10.7%	6.9%
TOTAL	100.0%	100.0%	100.0%

V=0.134**

Family income is indeed related to how happy respondents say they are (V = .13**), because people from high-income families are more likely (35.7 percent) than their low-income counterparts (22.4 percent) to report that they are "very happy." Conversely, those from low-income families are more likely (20.3 percent) than their high-income counterparts (6.9 percent) to say they're "not too happy."

One of the most important life chances is our health. The GSS asks respondents to rate their own health as excellent, good, fair, or poor.

Data File: **GSS**
Task: **Cross-tabulation**
➤ Row Variable: **133) HEALTH**
➤ Column Variable: **7) FAM INCOME**
➤ View: **Tables**
➤ Display: **Column %**

7) FAM INCOME			
	$1K-$19999	$20K-39999	$40K & UP
133) EXCELLENT	20.3%	31.5%	38.6%
GOOD	44.7%	51.3%	50.7%
FAIR/POOR	35.1%	17.3%	10.7%
TOTAL	100.0%	100.0%	100.0%

V=0.189**

Low-income people are less likely (20.3 percent) than high-income people (38.6 percent) to rate their health as excellent, and they are more likely to rate their health as only fair or poor (V = .19**). Social class certainly does seem to be related to our health.

Could the problems of low-income families lead to depression? The GSS asks how many days in the last week respondents felt that they "couldn't shake the blues." Let's see whether low-income people are more likely to feel this way.

Data File: **GSS**
Task: **Cross-tabulation**
➤ Row Variable: **140) SHAKE BLUE**
➤ Column Variable: **7) FAM INCOME**
➤ View: **Tables**
➤ Display: **Column %**

7) FAM INCOME			
	$1K-$19999	$20K-39999	$40K & UP
140) 0 DAYS	44.7%	55.9%	59.0%
1-2 DAYS	27.4%	28.0%	27.6%
3 OR MORE	27.9%	16.1%	13.5%
TOTAL	100.0%	100.0%	100.0%

V=0.116**

Low-income people are more likely (27.9 percent) than high-income people (13.5 percent) to "feel the blues" at least three days in the past week (V = .12**).

To move to a less serious but still important topic, let's now examine possible class differences in leisure-time activities.

How many hours a day do you watch TV? Have you ever wondered why some people watch more TV than others? Have you ever thought about why you watch the amount *you* do? Let's see whether TV watching varies by family income.

Data File: **GSS**
Task: **Cross-tabulation**
➤ Row Variable: **151) WATCH TV**
➤ Column Variable: **7) FAM INCOME**
➤ View: **Tables**
➤ Display: **Column %**

7) FAM INCOME

151) WATCH TV	$1K-$19999	$20K-39999	$40K & UP
0-1 HOURS	14.6%	19.6%	34.5%
2-3 HOURS	42.6%	50.5%	47.9%
4+ HOURS	42.8%	29.9%	17.6%
TOTAL	100.0%	100.0%	100.0%

V=0.187**

Family income is definitely related to TV watching. High-income respondents are more likely (34.5 percent) than low-income ones (14.6 percent) to watch only 0–1 hour per day. Conversely, low-income respondents are more likely to watch 4 or more hours per day (V = .19**).

Could family income affect how often people go out to the movies? If low-income people watch TV more often, perhaps they also go to the movies more often. On the other hand, that can get to be expensive, so perhaps they see fewer movies than people with higher incomes.

Data File: **GSS**
Task: **Cross-tabulation**
➤ Row Variable: **160) SEE FILM**
➤ Column Variable: **7) FAM INCOME**
➤ View: **Tables**
➤ Display: **Column %**

7) FAM INCOME

160) SEE FILM	$1K-$19999	$20K-39999	$40K & UP
YES	9.2%	13.8%	16.9%
NO	90.8%	86.2%	83.1%
TOTAL	100.0%	100.0%	100.0%

V=0.093*

Income seems to make a difference here, because the poorest respondents are almost half as likely than the wealthiest respondents not to have seen a movie in the past week (V = .09*). Of course, this table doesn't tell us why. Whether the poor simply can't afford the time or money to see a movie or, however unlikely, just don't like movie theaters must for now remain a mystery.

◆ EXERCISE 7 ◆
RACE AND ETHNICITY

Tasks: Mapping, Scatterplot, Univariate, Historical Trends, Cross-tabulation
Data Files: NATIONS, STATES, GSS, HISTORY

Perhaps the most important sociological aspect of race and ethnicity is that people of different races and ethnicities are treated differently. More precisely, some are treated better, while others are treated worse. The latter are said to belong to a racial or ethnic *minority group*, a category of people treated negatively because of their perceived physical or cultural characteristics. Negative outcomes for minority groups include housing and job discrimination, poverty and low education, and serious health problems. Minority groups in the United States and elsewhere have often been the victims of hate crimes, and in other nations have suffered virtual genocide.

This chapter explores several aspects of race and ethnicity in the United States and throughout the world. Our major focus will be on the extent and correlates of racial and ethnic prejudice and on the negative life chances of minority groups. We'll begin by exploring these topics on the international level and then switch our attention to the United States.

RACE AND ETHNICITY IN THE INTERNATIONAL ARENA

It's surprisingly difficult to measure how much racial and ethnic diversity exists within any single nation. Using a complex formula, sociologist Rodney Stark calculated the odds that any two persons in a given nation will differ in their race, religion, ethnicity or tribe, or language group. Let's use this measure of multiculturalism to see which regions of the world have the most and least diversity.

> *Data File:* **NATIONS**
> *Task:* **Mapping**
> *Variable 1:* **55) MULTI-CULT**
> *View:* **Map**

MULTI-CULT -- MULTI-CULTURALISM:ODDS THAT ANY 2 PERSONS WILL DIFFER IN THEIR RACE, RELIGION, ETHNICITY (TRIBE),OR LANGUAGE GROUP (STARK)

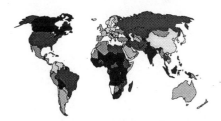

The lighter the color, the less diversity. Despite several exceptions, overall the least diversity appears in parts of Europe, Asia, and Central America, while the highest diversity appears in Africa.

In the surveys included in the NATIONS data set, respondents were asked four questions about their willingness to live, respectively, near Jews, foreigners, Muslims, and members of another race. The NATIONS data set reports the percent of each nation's respondents who said they would *not* want members of these groups to be their neighbors. These variables indicate the extent of the nations' racial and ethnic prejudice. Let's take a brief moment to map each of these variables (you'll also find it interesting to rank the results of each map).

Data File: **NATIONS**
Task: **Mapping**
➤ Variable 1: **58) ANTI-SEM.**
➤ View: **Map**

ANTI-SEM. -- PERCENT WHO WOULD NOT WANT JEWS AS NEIGHBORS (WVS)

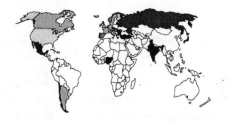

Data File: **NATIONS**
Task: **Mapping**
➤ Variable 1: **59) ANTI-FORGN**
➤ View: **Map**

ANTI-FORGN -- PERCENT WHO WOULD NOT WANT FOREIGNERS AS NEIGHBORS (WVS)

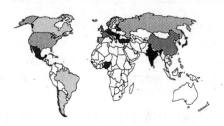

Data File: **NATIONS**
Task: **Mapping**
➤ Variable 1: **60) ANTI-MUSLM**
➤ View: **Map**

ANTI-MUSLM -- PERCENT WHO WOULD NOT WANT MUSLIMS AS NEIGHBORS (WVS)

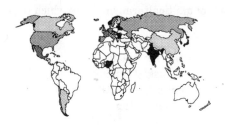

Data File: **NATIONS**
Task: **Mapping**
➤ Variable 1: **61) RACISM**
➤ View: **Map**

RACISM -- PERCENT WHO WOULD NOT WANT MEMBERS OF ANOTHER RACE AS NEIGHBORS (WVS)

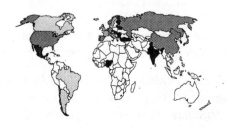

Discovering Sociology

Generally, all four maps look very similar. The most prejudiced nations are generally found in Eastern Europe and parts of Asia.

Why are the people of some nations more likely than those in other nations to be prejudiced? Maybe education makes a difference. As people acquire a formal education, they tend to learn about other societies and cultures and begin to appreciate the lives of people with different racial and ethnic backgrounds from their own. If this is true, the nations with the highest levels of education should be least prejudiced. Let's examine this possibility with the variable measuring the percent of the population that would not want Jews as neighbors.

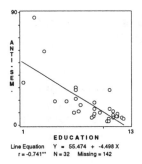

        Data File: **NATIONS**
        Task: **Scatterplot**
➤ *Dependent Variable:* **58) ANTI-SEM.**
➤ *Independent Variable:* **40) EDUCATION**
        ➤ *Display:* **Reg. Line**

The higher a nation's education, the lower its degree of anti-Semitism (r = –.74**).

RACE AND ETHNICITY IN THE UNITED STATES

We now turn our attention to the United States. Let's first get a picture of the U.S. racial and ethnic composition using Census data. As we do, see where your home state fits in.

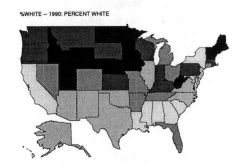

    ➤ *Data File:* **STATES**
      ➤ *Task:* **Mapping**
  ➤ *Variable 1:* **13) %WHITE**
      ➤ *View:* **Map**

This map uses U.S. Census data to indicate the percent of each state's population that is white; the darker the color, the higher the percent white. As you can see, this percent tends to be highest in the upper Midwest and in upper New England, and lowest in the South.

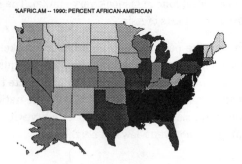

Data File: **STATES**
Task: **Mapping**
➤ *Variable 1:* **14) %AFRIC.AM**
➤ *View:* **Map**

The highest proportion of African Americans live in the Southeast and, more generally, east of the Mississippi River.

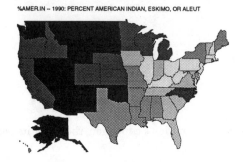

Data File: **STATES**
Task: **Mapping**
➤ *Variable 1:* **15) %AMER.IN**
➤ *View:* **Map**

The states that have the highest proportion of Native Americans lie west of the Mississippi River.

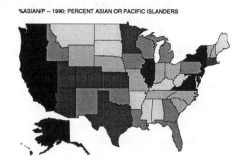

Data File: **STATES**
Task: **Mapping**
➤ *Variable 1:* **16) %ASIAN/P**
➤ *View:* **Map**

The two major regions for Asians and Pacific Islanders are in the Far West and in the Northeast.

Data File: **STATES**
Task: **Mapping**
➤ *Variable 1:* **23) %HISPANIC**
➤ *View:* **Map**

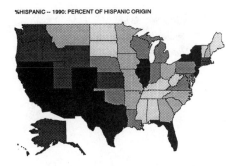

The Southwest and some states east of the Mississippi have the highest proportion of people of Hispanic origin.

We'll now switch to the GSS and examine the extent and predictors of various views on racial issues.

The GSS includes several items measuring such views. One item asked whether there should be laws against marriages between blacks and whites.

➤ *Data File:* **GSS**
➤ *Task:* **Univariate**
➤ *Primary Variable:* **26) INTERMAR.?**
➤ *View:* **Pie**

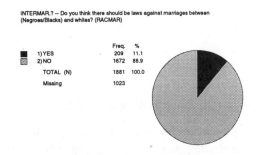

About 11 percent of the U.S. population believes there should be laws prohibiting racial intermarriages. Has this percent declined in the past 25 years?

➤ *Data File:* **HISTORY**
➤ *Task:* **Historical Trends**
➤ *Variable:* **5) INTERMAR?**

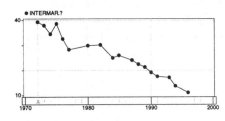

Percent agreeing with laws against racial intermarriage

The percent of the U.S. population favoring laws prohibiting racial intermarriages has declined considerably since the early 1970s.

A second variable asks respondents whether they agree that "white people have the right to keep blacks out of their neighborhoods if they want to, and blacks should respect that right."

➤ *Data File:* **GSS**
➤ *Task:* **Univariate**
➤ *Primary Variable:* **27) RACE SEG.**
➤ *View:* **Pie**

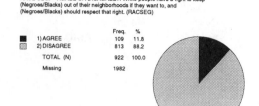

About 12 percent of the population believes in the right of whites to practice housing segregation.

➤ *Data File:* **HISTORY**
➤ *Task:* **Historical Trends**
➤ *Variable:* **7) RACE SEG**

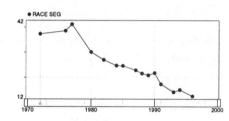

Percent favoring racial segregation in housing

This percent has also dropped considerably since the early 1970s.

Our analysis will now focus on white people, and our dependent variable will be another GSS question, "How would you respond to a close relative marrying a black person?" We'll first see how whites respond to this question by using ExplorIt's subset option.

➤ *Data File:* **GSS**
➤ *Task:* **Univariate**
➤ *Primary Variable:* **37) MARRY BLK**
➤ *Subset Variable:* **42) WHTE/AFRAM**
➤ *Subset Categories:* **Include: 1) White**
➤ *View:* **Pie**

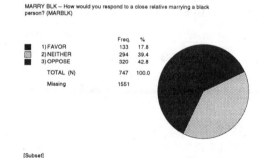

[Subset]

The option for selecting a subset variable is located on the same screen you use to select other variables. For this example, select 42) WHITE/AFRAM as a subset variable. A window will appear that shows you the categories of the subset variable. Select 1) White as your subset category and choose the [Include] option. Then click [OK] and continue as usual. With this particular subset selected, the results will be limited to the WHITES in the sample.

Discovering Sociology

About 43 percent of whites would oppose a close relative's marrying an African American.

Historically, the South has been more prejudiced than other U.S. regions. Are white Southerners today more likely than non-Southerners to oppose a relative's marrying an African American?

Data File:	**GSS**			
➤ Task:	**Cross-tabulation**			
➤ Row Variable:	**37) MARRY BLK**			
➤ Column Variable:	**89) SOUTH**			
➤ Subset Variable:	**42) WHTE/AFRAM**			
➤ Subset Categories:	**Include: 1) White**			
➤ View:	**Tables**			
➤ Display:	**Column %**			

89) SOUTH		
	SOUTH	NON-SOUT
37) FAVOR	10.0%	21.9%
NEITHER	32.8%	42.8%
OPPOSE	57.1%	35.2%
TOTAL	100.0%	100.0%

V=0.222**

White Southerners are more likely (57.1 percent) than their non-southern counterparts (35.2 percent) to oppose a close relative's marrying an African American (V = .22**).

Earlier we saw that in the international arena, higher levels of education were linked to lower opposition to living near Jews. If this relationship holds true within the United States, then, among whites, Americans without a high school degree should be more likely than those with a college degree to oppose a close relative's marrying an African American.

Data File:	**GSS**
Task:	**Cross-tabulation**
Row Variable:	**37) MARRY BLK**
➤ Column Variable:	**6) EDUCATION**
➤ Subset Variable:	**42) WHTE/AFRAM**
➤ Subset Categories:	**Include: 1) White**
➤ View:	**Tables**
➤ Display:	**Column %**

6) EDUCATION				
	NO HS GRA	HS GRAD	SOME COL	COLL GRA
37) FAVOR	16.1%	18.2%	15.2%	21.2%
NEITHER	27.1%	35.4%	37.6%	51.9%
OPPOSE	56.8%	46.4%	47.1%	26.9%
TOTAL	100.0%	100.0%	100.0%	100.0%

V=0.154**

The subset from the above example is still restricted to 1) Whites unless you have exited the task, deleted the subset variables, or cleared all variables.

Whites without a high school degree are more likely than (56.8 percent) those with a college degree (26.9 percent) to oppose a close relative's marrying an African American (V = .15**).

In the United States, some whites live in white-only neighborhoods, while other whites live in integrated neighborhoods. If "familiarity breeds contempt," then whites in integrated neighborhoods should be *more* likely than those in segregated neighborhoods to oppose a close relative's marrying an African American. If, on the other hand, whites in integrated neighborhoods get to know their African-American neighbors as real people, and not just as stereotypes, then they should be *less* likely than those in segregated neighborhoods to oppose a close relative's marrying an African American. Let's see whether either hypothesis receives support. Our independent variable will be a GSS question, "Are there any blacks living in this neighborhood right now?"

Data File: **GSS**
Task: **Cross-tabulation**
Row Variable: **37) MARRY BLK**
➤ Column Variable: **28) BL.IN AREA**
➤ Subset Variable: **42) WHTE/AFRAM**
➤ Subset Categories: **Include: 1) White**
➤ View: **Tables**
➤ Display: **Column %**

28) BL.IN AREA

		YES	NO
37)	FAVOR	20.8%	14.4%
MARRY	NEITHER	40.7%	34.4%
BLK	OPPOSE	38.6%	51.1%
	TOTAL	100.0%	100.0%

V=0.127**

Again, the subset is still restricted to 1) Whites unless you have exited the task, deleted the subset variables, or cleared all variables.

Our second hypothesis is supported: whites in integrated neighborhoods are *less* likely (38.6 percent) than those in segregated neighborhoods (51.1 percent) to oppose a close relative's marrying an African American (V = .13**). Our brief exploration suggests that whites' opposition to a close relative's marrying an African American is highest among Southerners, among people without a high school degree, and among people living in segregated neighborhoods.

◆ EXERCISE 8 ◆
GENDER AND GENDER INEQUALITY

Tasks: Mapping, Scatterplot, Univariate, Historical Trends, Cross-tabulation
Data Files: NATIONS, STATES, GSS, HISTORY

The different roles expected of females and males are our ***genders***. Whereas sex is a biological category determined at the moment of conception, gender is a social and cultural one. Therefore, gender and gender roles differ from one society to another. Even within a given society, various people have different expectations of how girls and boys and women and men should think and behave.

If, according to the sociological perspective, our social backgrounds influence our behavior, attitudes, and life chances, then gender illustrates the sociological perspective perhaps more than any other social category. Simply put, gender has a profound influence on many aspects of how we think, of how we act, and of our life chances. ***Gender socialization*** refers to gender's influence on how we think and act, while ***gender inequality*** refers to gender's influence on our life chances. In speaking of gender inequality, sociologists emphasize that men are the dominant sex and women the subordinate sex. When it comes to such things as wealth, power, prestige, and other life outcomes, men often fare much better than women. To acknowledge such gender inequality is not to suggest there has to be such inequality. Rather, an understanding of the sources and manifestations of gender inequality is crucial if we are to achieve a society in which women and men are equal.

The data sets included with this workbook have many variables related to gender, far too many for us to use all of them. We'll look at just a few, and begin by trying to get some idea of international differences in gender inequality and in views on some important gender issues. Then we'll turn to the United States to see what difference being female or male makes in behavior, attitudes, and life chances. Worksheets will give you the opportunity to explore some gender topics on your own.

CROSS-CULTURAL DIFFERENCES IN GENDER AND GENDER INEQUALITY

People in nations around the world differ in their views about gender issues and in their degree of gender equality. Your NATIONS data set includes a variable listing the percentage of each country's population who agree that "what women really want is a home and children." Whether or not you agree with this belief, it obviously represents a traditional view of women's gender role. Let's see how responses to this question differ across the world.

➤ *Data File:* **NATIONS**
➤ *Task:* **Mapping**
➤ *Variable 1:* **80) HOME&KIDS**
➤ *View:* **Map**

HOME&KIDS -- PERCENT WHO AGREE THAT WHAT "WOMEN REALLY WANT IS A HOME AND CHILDREN" (WVS)

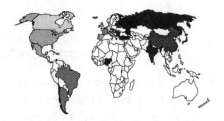

The darker the color, the higher the percent who feel that women want a home and children most of all. The nations that have the highest percent of residents who feel this way tend to be in Eastern Europe and the underdeveloped world, while the nations whose residents take a more contemporary view—those lighter in color—tend to be in Western Europe and North America.

Data File: **NATIONS**
Task: **Mapping**
Variable 1: **80) HOME&KIDS**
➤ *View:* **List: Rank**

RANK	CASE NAME	VALUE
1	Lithuania	97.00
2	India	94.00
3	Czech Republic	93.00
3	Slovak Republic	93.00
5	Bulgaria	90.00
5	Latvia	90.00
5	Russia	90.00
8	Turkey	88.00
8	Nigeria	88.00
10	Estonia	85.00

The variation in responses to this question is remarkable. At the high end, 97 percent of Lithuanians and 94 percent of the people of India think that women want a home and children most of all. At the low end, only 25 percent of Denmark's residents feel this way. Note that although the United States is near the low end on this question, more than half of U.S. respondents, or 56 percent, still think that women want a home and children above all. Canada stands 13 percent lower, at 43 percent. Thus the United States is a bit more traditional than Canada on this one aspect of women's gender role.

Now let's look at international variation in gender inequality. Although there are many ways of measuring this concept, we'll use a variable that lists the average female years of schooling as a percentage of the male years of schooling. Thus the lower the percentage, the less education women have compared to men.

Data File: **NATIONS**
➤ *Task:* **Mapping**
➤ *Variable 1:* **74) M/F EDUC**
➤ *View:* **Map**

M/F EDUC. -- AVERAGE FEMALE YEARS OF SCHOOLING AS A PERCENTAGE OF AVERAGE MALE YEARS (CALCULATED)

The lighter the color, the less education women have compared to men. Women's education lags farthest behind men's in the underdeveloped world, primarily in the nations of Africa and parts of Asia.

Do beliefs about gender roles vary with gender equality? It makes sense to think that the most traditional beliefs are found in the nations with the least gender equality. Let's see whether this is true by using our previous attitudinal measure on women's role in the home along with an overall measure of gender equality that takes into account several aspects of women's lives.

Data File: **NATIONS**
➤ *Task:* **Scatterplot**
➤ *Dependent Variable:* **80) HOME&KIDS**
➤ *Independent Variable:* **75) GENDER EQ**
➤ *View:* **Reg. Line**

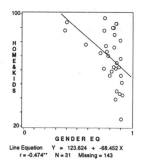

Line Equation   Y = 123.624 + -68.452 X
r = -0.474**   N = 31   Missing = 143

We see the relationship we expected. People in the nations with the least gender equality are more likely to believe that women want a home and children above all. The correlation is a relatively high −.47.

GENDER AND GENDER INEQUALITY IN THE UNITED STATES

We'll begin our U.S. focus on gender by looking at the STATES data set. We know that, biologically, there's a 50 percent chance that a baby will be a girl and a 50 percent chance that it will be a boy. Thus about half of all babies are female and half are male. However, since men on the average die sooner than women, a little more than 50 percent of all people are female. This percent varies slightly by state because of migration patterns. Let's look at the map for the percent of each state that is female.

> ➤ *Data File:* **STATES**
> ➤ *Task:* **Mapping**
> ➤ *Variable 1:* **10) %FEMALE**
> ➤ *View:* **Map**

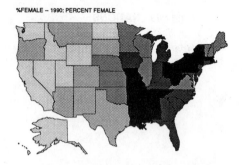

The states east of the Mississippi River have the highest percentages of female residents, and the states west of the Mississippi have the lowest percentages. What might explain this geographic pattern?

One measure of gender equality is the percent of legislators who are women. The STATES data set includes such a percent for state legislators.

> *Data File:* **STATES**
> *Task:* **Mapping**
> ➤ *Variable 1:* **115) %FEMALE LG**
> ➤ *View:* **Map**

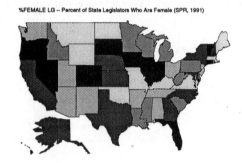

No clear geographic pattern emerges here. Although some states have higher proportions of women in their legislatures, these states seem to be scattered throughout the country.

> *Data File:* **STATES**
> *Task:* **Mapping**
> *Variable 1:* **115) %FEMALE LG**
> ➤ *View:* **List: Rank**

RANK	CASE NAME	VALUE
1	Vermont	34.4
2	South Dakota	32.8
3	Florida	31.8
4	Nevada	31.7
5	Oregon	31.3
6	Connecticut	31.0
7	Illinois	28.6
8	Indiana	27.6
9	Iowa	26.7
10	Colorado	24.8

Six states, led by Vermont, have more than 30 percent female membership in their state legislatures. At the other end, only 2.8 percent of North Dakota's state legislators are women. Is it fair to say that women are closer to equality with men in the states where they hold higher proportions of legislative seats?

Let's turn to the GSS and begin with some views on women's roles. We'll first examine the percent of respondents who agree that "women should take care of running their homes and leave running the country up to men." Do you agree or disagree with this statement?

➤ *Data File:* **GSS**
    ➤ *Task:* **Univariate**
➤ *Primary Variable:* **46) WOMEN HOME**
    ➤ *View:* **Pie**

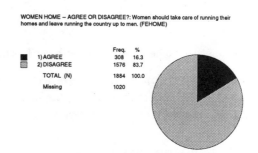

WOMEN HOME -- AGREE OR DISAGREE?: Women should take care of running their homes and leave running the country up to men. (FEHOME)

		Freq.	%
■	1) AGREE	308	16.3
▨	2) DISAGREE	1576	83.7
	TOTAL (N)	1884	100.0
	Missing	1020	

About 16 percent of the U.S. population agrees that women should take care of running their homes and leave running the country up to men.

Another GSS item points out that women are more likely than men to take care of children and asks how important the following reason is for this situation: "It is God's will that women care for children."

Data File: **GSS**
    Task: **Univariate**
➤ *Primary Variable:* **51) CARE CHLD5**
    ➤ *View:* **Pie**

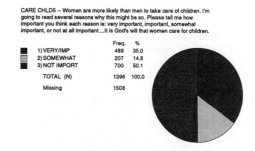

CARE CHLD5 -- Women are more likely than men to take care of children. I'm going to read several reasons why this might be so. Please tell me how important you think each reason is: very important, important, somewhat important, or not at all important....It is God's will that women care for children.

		Freq.	%
■	1) VERY/IMP	489	35.0
▨	2) SOMEWHAT	207	14.8
■	3) NOT IMPORT	700	50.1
	TOTAL (N)	1396	100.0
	Missing	1508	

About half the sample think that God's will is a very important or somewhat important reason for the fact that women are more likely than men to take care of children.

These two questions both touch on women's traditional home and family role and together indicate that a significant minority of the public does continue to accept this traditional role. Has this acceptance declined in the last quarter century? Let's look at the trend for responses to the first question, which that asked whether women should take care of running their homes.

> ➤ *Data File:* **HISTORY**
>   ➤ *Task:* **Historical Trends**
>   ➤ *Variable:* **10) WOMEN HOME**

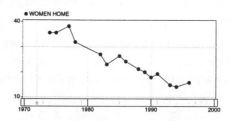

Percent saying women should take care of running their homes

The belief that women should stay at home has declined considerably in the last quarter century.

One of the most important consequences of gender for life chances lies in the workplace, where women's earnings continue to lag far behind men's. To illustrate this, let's look at the annual earnings—combined into three categories—of male and female workers.

> ➤ *Data File:* **GSS**
>   ➤ *Task:* **Cross-tabulation**
>   ➤ *Row Variable:* **8) OWN INCOME**
> ➤ *Column Variable:* **45) GENDER**
> ➤ *Subset Variable:* **1) WORKING?**
> ➤ *Subset Categories:* **Include: 1) Working**
>   ➤ *View:* **Tables**
>   ➤ *Display:* **Column %**

	45) GENDER	
	FEMALE	MALE
8) $1K-$19999	52.6%	25.6%
$20K-39999	33.9%	41.9%
$40K & UP	13.5%	32.5%
TOTAL	100.0%	100.0%

V=0.301**

The option for selecting a subset variable is located on the same screen you use to select other variables. For this example, select 1) WORKING? as a subset variable. A window will appear that shows you the categories of the subset variable. Select 1) Working as your subset category and choose the [Include] option. Then click [OK] and continue as usual.

Among women, 52.6 percent indicate that their personal income is in the lowest income bracket. Among men, this percentage was only 25.6 percent. Conversely, men are more likely (32.5 percent) than women (13.5 percent) to be in the highest bracket (V = .30**). But, does this difference persist when we look only at people who have college degrees? Today most college students, male or female, want to get a degree to increase their future earnings. Let's see whether gender affects the income we can expect to receive after we graduate from college. We'll have to use two subset variables this time.

> *Data File:* **GSS**
> *Task:* **Cross-tabulation**
> *Row Variable:* **8) OWN INCOME**
> *Column Variable:* **45) GENDER**
> *Subset Variable 1:* **1) WORKING?**
> *Subset Categories:* **Include: 1) Working**
> ➤ *Subset Variable 2:* **6) EDUCATION**
> ➤ *Subset Categories:* **Include: 4) College Grad**
>   ➤ *View:* **Tables**
>   ➤ *Display:* **Column %**

	45) GENDER	
	FEMALE	MALE
8) $1K-$19999	30.1%	11.5%
$20K-39999	41.5%	32.0%
$40K & UP	28.4%	56.5%
TOTAL	100.0%	100.0%

V=0.307**

Discovering Sociology

To add a subset variable, return to the same screen you use to select other variables. For this example, select 6) EDUCATION as a second subset variable. A window will appear that shows you the categories of the subset variable. Select 4) College Grad as your subset category and choose the [Include] option. Both 6) EDUCATION and 1) WORKING? are now listed as subset variables. Then click [OK] and continue as usual.

Even though a greater percent of both sexes with B.A. degrees have high incomes than was true in the previous example for the whole sample, women continue to lag far behind men. They are more likely to be in the lowest income bracket (about a 19 percentage point difference) and less likely to be in the highest bracket (about a 28 percentage point difference). The relationship is strong and statistically significant (V = .31**). Why do you think these income differences exist? Note, by the way, that the 28 percent gender difference for high incomes among people with college degrees is even larger than the 19 percent difference for high incomes we saw above for the whole sample.

Another important gender issue today is the women's movement itself, on which people continue to have strong views. The GSS asked its respondents whether or not the women's movement has improved their lives.

    Data File: **GSS**
   ➤ *Task:* **Univariate**
➤ *Primary Variable:* **53) YOURSELF**
   ➤ *View:* **Pie**

YOURSELF -- DOES RESPONDENT THINK THE WOMEN'S MOVEMENT HAS IMPROVED HIS OR HER LIFE OR NOT? (YOURSELF)

	Freq.	%
■ 1) IMPROVED	546	38.8
▨ 2) NOT IMPROV	861	61.2
TOTAL (N)	1407	100.0
Missing	1497	

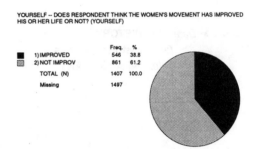

About 39 percent feel the women's movement has improved their lives.

Do women and men have different views as to whether the women's movement has helped them? Let's find out. What do you predict?

    Data File: **GSS**
   ➤ *Task:* **Cross-tabulation**
  ➤ *Row Variable:* **53) YOURSELF**
➤ *Column Variable:* **45) GENDER**
   ➤ *View:* **Tables**
  ➤ *Display:* **Column %**

		45) GENDER	
		FEMALE	MALE
53) YOURSELF	IMPROVED	47.2%	28.0%
	NOT IMPROV	52.8%	72.0%
	TOTAL	100.0%	100.0%

V=0.196**

Among women, 47.2 percent think that the women's movement has improved their lives. Among men this percentage was only 28.0 percent (V = .20**).

Finally, let's turn to one of the most controversial gender-related issues today, abortion. The GSS asks whether a woman should be allowed to have a legal abortion if she wants one "for any reason."

Data File: **GSS**
➤ Task: **Univariate**
➤ Primary Variable: **68) ABORT ANY**
➤ View: **Pie**

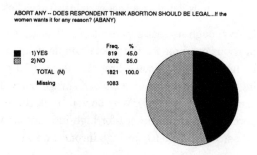

ABORT ANY -- DOES RESPONDENT THINK ABORTION SHOULD BE LEGAL...if the women wants it for any reason? (ABANY)

		Freq.	%
■	1) YES	819	45.0
▨	2) NO	1002	55.0
	TOTAL (N)	1821	100.0
	Missing	1083	

Forty-five percent of the U.S. population thinks a woman should be allowed to have a legal abortion for any reason.

➤ Data File: **HISTORY**
➤ Task: **Historical Trends**
➤ Variable: **24) ABORT ANY**

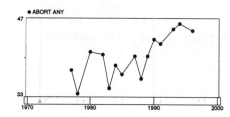

Percent favoring legal abortion for any reason

This figure has risen overall since the 1970s.

Do you think gender affects support for legalized abortion? It would probably make sense to think that women should support it more than men. Let's find out.

➤ Data File: **GSS**
➤ Task: **Cross-tabulation**
➤ Row Variable: **68) ABORT ANY**
➤ Column Variable: **45) GENDER**
➤ View: **Tables**
➤ Display: **Column %**

		45) GENDER	
		FEMALE	MALE
68)	YES	45.5%	44.4%
ABORT ANY	NO	54.5%	55.6%
	TOTAL	100.0%	100.0%

V=0.011

Surprise! Women are no more likely than men to support legalized abortion.

Discovering Sociology

AGE AND AGING

Tasks: Mapping, Univariate, Scatterplot, Cross-tabulation, Historical Trends
Data Files: NATIONS, STATES, GSS, HISTORY

A s we get older, society's view of us changes. TV shows and commercials feature the young and extol the excitement of youth, while both often portray the elderly as forgetful and frail and even as buffoons. Our society's emphasis on youthfulness leads many older Americans to try to prevent or minimize the physiological effects of aging with things like wrinkle creams and plastic surgery. But time marches on, and eventually many do become frail and see their spouses, friends, and acquaintances develop health problems and, sometimes, die. Many elderly people, especially women whose husbands have already passed away, live by themselves. It's certainly not an easy time of life, but neither is it as lamentable as it's often portrayed in the media. The elderly often lead happy and healthy lives and certainly don't fit the stereotypes that characterize them in the United States and many other nations.

Your workbook's data sets include several variables related to age and aging that we will explore in the next few pages. We'll look at the geographic distribution of older people and at the difference, if any, that age makes in various attitudes, behaviors, and life chances.

AGE AND AGING IN INTERNATIONAL PERSPECTIVE

Life expectancy differs dramatically around the world. Let's see which nations have the longest and shortest expected life spans.

> *Data File:* **NATIONS**
> *Task:* **Mapping**
> *Variable 1:* **19) LIFE EXPCT**
> *View:* **Map**

LIFE EXPCT -- AVERAGE LIFE EXPECTANCY (TWF 1994)

The lighter the color, the lower the life expectancy. The nations of Africa and much of Asia tend to have the lowest life spans, the nations of North America and Europe the highest.

Euthanasia, or terminating the life of someone with a fatal medical condition, is one of the more controversial issues today. It especially affects the elderly: while the young can have fatal medical conditions, the old are much more likely to suffer from them. Views on euthanasia differ widely around the world. The NATIONS data set includes a variable listing the percent of each nation's residents who believe euthanasia is acceptable.

Data File: **NATIONS**
Task: **Mapping**
➤ Variable 1: **116) EUTHANASIA**
➤ View: **Map**

EUTHANASIA – PERCENT WHO BELIEVE EUTHANASIA IS OK (TERMINATING THE LIFE OF THE INCURABLY SICK) (WVS)

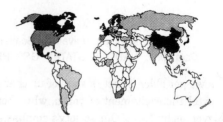

The pattern in this map isn't as clear as the patterns in other maps we've been seeing, where there have been clear differences between the developed and underdeveloped nations. In particular, the European nations seem to have divergent views on euthanasia; some are very likely to think it's acceptable, while others are much more disapproving.

What explains why some nations are more likely than others to approve of euthanasia? Many variables might come to mind, but let's look at religiosity—how religious people are. Does it makes sense to think that the more religious a nation's people, the less likely they are to approve of euthanasia? Let's find out.

Data File: **NATIONS**
➤ Task: **Scatterplot**
➤ Dependent Variable: **116) EUTHANASIA**
➤ Independent Variable: **51) CH.ATTEND**
➤ View: **Reg. Line**

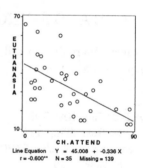

Line Equation   Y = 45.008 + -0.336 X
r = -0.600**   N = 35   Missing = 139

Our hypothesis is supported: the more frequent the religious attendance, the lower the support for euthanasia. The correlation, r, is a high –.60.

AGE AND AGING IN THE UNITED STATES

About 12.5 percent of the U.S. population is 65 and older. Yet this percentage varies across the country. Some states have higher proportions of the elderly, while others have lower proportions.

> ➤ *Data File:* **STATES**
> ➤ *Task:* **Mapping**
> ➤ *Variable 1:* **5) %>AGE 64**
> ➤ *View:* **Map**

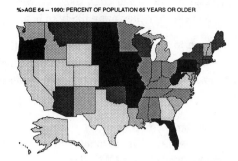

%>AGE 64 -- 1990: PERCENT OF POPULATION 65 YEARS OR OLDER

The states with the highest proportions of the elderly tend to be in the Midwest, but four states in the East also are in the top tier. The West generally has the lowest percent of elderly people. Why do you think this is so?

We can also see what percent of each state's population lives in nursing homes.

> *Data File:* **STATES**
> *Task:* **Mapping**
> ➤ *Variable 1:* **51) %NURS.HOME**
> ➤ *View:* **Map**

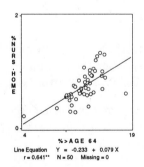

%NURS.HOME -- 1990: PERCENT LIVING IN NURSING HOMES

Not surprisingly, this map looks similar to the one for percent elderly.

> *Data File:* **STATES**
> ➤ *Task:* **Scatterplot**
> ➤ *Dependent Variable:* **51) %NURS.HOME**
> ➤ *Independent Variable:* **5) %>AGE 64**
> ➤ *View:* **Reg. Line**

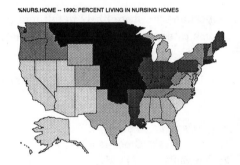

Line Equation   Y = -0.233 + 0.079 X
r = 0.641**   N = 50   Missing = 0

The scatterplot confirms that the percent who are elderly in a state is strongly related (r = .64**) to the state's percent in nursing homes. One puzzle that arises from inspecting the maps is why Florida's nursing home percent is not in the top tier, given that its percent who are elderly is so high. Are its

Exercise 9: Age and Aging   123

elderly residents healthier than those elsewhere? Wealthier (and hence able to afford better care than nursing homes)? Are they more likely to have their children living near them? What do you think?

We now turn to the GSS to see how, if at all, the elderly differ from younger age groups. We often hear that the elderly, many of whom are on fixed incomes, are especially likely to have financial problems. Let's see whether age is related to family income from all sources.

➤ *Data File:* **GSS**
   ➤ *Task:* **Cross-tabulation**
➤ *Row Variable:* **7) FAM INCOME**
➤ *Column Variable:* **90) AGE 65+**
   ➤ *View:* **Tables**
   ➤ *Display:* **Column %**

	90) AGE 65+	
7) FAM INCOME	18-64	65+
$1K-$19999	24.9%	51.8%
$20K-39999	31.6%	32.1%
$40K & UP	43.5%	16.2%
TOTAL	100.0%	100.0%

V=0.225**

Older Americans are more likely to fall into the low-income category, and less likely to fall into the high-income category (V = .23**). On the whole, they are indeed poorer than their younger counterparts.

Does the elderly's worse financial situation translate into dissatisfaction with their finances?

Data File: **GSS**
Task: **Cross-tabulation**
➤ *Row Variable:* **13) SAT.$?**
➤ *Column Variable:* **90) AGE 65+**
   ➤ *View:* **Tables**
   ➤ *Display:* **Column %**

	90) AGE 65+	
13) SAT.$?	18-64	65+
PRETTY WEL	24.5%	45.9%
MORE/LESS	45.8%	37.1%
NOT SATIS.	29.8%	17.0%
TOTAL	100.0%	100.0%

V=0.177**

Far from it! Those over 65 are *more* likely (45.9 percent) to say they're "pretty well" satisfied with their financial situation than are those under 65 (24.5 percent). The elderly are also less likely to say they're "not satisfied" (V = .18**). How would you explain this surprising result?

Another common perception of the elderly is that they spend many days by themselves and thus are very lonely. Let's see whether the GSS data confirm this. We'll first see whether the elderly are more likely to live alone.

<table>
<tr><td>Data File:</td><td>GSS</td></tr>
<tr><td>Task:</td><td>Cross-tabulation</td></tr>
<tr><td>➤ Row Variable:</td><td>61) ALONE?</td></tr>
<tr><td>➤ Column Variable:</td><td>90) AGE 65+</td></tr>
<tr><td>➤ View:</td><td>Tables</td></tr>
<tr><td>➤ Display:</td><td>Column %</td></tr>
</table>

	90) AGE 65+	
	18-64	65+
61) ALONE		
ALONE	20.7%	52.3%
NOT ALONE	79.3%	47.7%
TOTAL	100.0%	100.0%

V=0.261**

The elderly are more likely (52.3 percent) to live alone than are younger people (20.7 percent).

Does this mean the elderly will feel more lonely? Our dependent variable will be the number of days in the last week on which the respondent felt lonely.

<table>
<tr><td>Data File:</td><td>GSS</td></tr>
<tr><td>Task:</td><td>Cross-tabulation</td></tr>
<tr><td>➤ Row Variable:</td><td>141) LONELY</td></tr>
<tr><td>➤ Column Variable:</td><td>90) AGE 65+</td></tr>
<tr><td>➤ View:</td><td>Tables</td></tr>
<tr><td>➤ Display:</td><td>Column %</td></tr>
</table>

	90) AGE 65+	
	18-64	65+
141) LONELY		
0 DAYS	53.8%	56.1%
1-2 DAYS	22.9%	19.5%
3 OR MORE	23.4%	24.4%
TOTAL	100.0%	100.0%

V=0.030

Another surprise: older Americans are no more likely than their younger counterparts to feel lonely (V = .03)!

What about general happiness? Would you predict that the elderly should feel more happy than their younger counterparts, less happy, or about the same? (You might not want to predict anything at this point!) Let's find out.

<table>
<tr><td>Data File:</td><td>GSS</td></tr>
<tr><td>Task:</td><td>Cross-tabulation</td></tr>
<tr><td>➤ Row Variable:</td><td>132) HAPPY?</td></tr>
<tr><td>➤ Column Variable:</td><td>90) AGE 65+</td></tr>
<tr><td>➤ View:</td><td>Tables</td></tr>
<tr><td>➤ Display:</td><td>Column %</td></tr>
</table>

	90) AGE 65+	
	18-64	65+
132) HAPPY?		
VERY HAPPY	29.4%	36.0%
PRET.HAPPY	58.6%	51.6%
NOT TOO	12.0%	12.4%
TOTAL	100.0%	100.0%

V=0.054*

The elderly are *more* likely (36.0 percent) than younger people (29.4 percent) to say that they are very happy (V = .05*). The difference is slight, but certainly doesn't fit the common assumption that the elderly lead despondent lives.

Still another perception of the elderly is that they suffer from senility and other problems having to do with thinking processes. Although the GSS doesn't have any direct measures of these problems, it does include a vocabulary test of 10 words, some of them easy, others difficult. Presumably, if the elderly do suffer from these problems, they should fare much worse on this test than their younger counterparts. Let's look at the number of correct answers for each group.

		90) AGE 65+	
		18-64	65+
146) #CRCT.WORD	0-5	36.6%	44.2%
	6	22.5%	21.2%
	7 OR MORE	40.9%	34.5%
	TOTAL	100.0%	100.0%

Data File: **GSS**
Task: **Cross-tabulation**
➤ Row Variable: **146) #CRCT.WORD**
➤ Column Variable: **90) AGE 65+**
➤ View: **Tables**
➤ Display: **Column %**

V=0.058*

Older Americans do have lower scores, but the difference is very slight (V = .06*). Moreover, it might be due to their lower educational levels rather than to senility or other such problems. Let's test this assumption by using education as a *control* variable. A control variable is different than a subset variable. As you'll recall, a subset variable allows you to select which survey respondents on a third variable you want to include (or exclude) from your analysis. A control variable, which we'll use here, doesn't simply exclude cases, but instead, breaks down the analysis so that the cases for each category of the control variable are shown as separate tables. In this example, we'll repeat the above analysis of #CRCT.WORD and AGE65+ but will see separate tables for each category of the EDUCATION variable. If the elderly are not more likely than younger respondents to have lower scores once education is taken into account, then within each education group the elderly should no longer have lower scores.

[Control: NO HS GRAD]

Data File: **GSS**
Task: **Cross-tabulation**
➤ Row Variable: **146) #CRCT.WORD**
➤ Column Variable: **90) AGE 65+**
➤ Control Variable: **6) EDUCATION**
➤ View: **Tables (NO HS GRAD)**
➤ Display: **Column %**

		90) AGE 65+	
		18-64	65+
146) #CRCT.WORD	0-5	72.9%	74.0%
	6	17.2%	15.6%
	7 OR MORE	9.9%	10.4%
	TOTAL	100.0%	100.0%

V=0.021

The option for selecting a control variable is located on the same screen you use to select other variables. For this example, select 6) EDUCATION as a control variable and then click [OK] to continue as usual. Separate tables for each of the 6) EDUCATION categories will now be shown for the 146) #CRCT.WORD and 90) AGE 65+ cross-tabulation

The above table includes only those respondents who don't have a high school degree. Compare the results for the two age groups before continuing.

Discovering Sociology

➤ *View:* **Tables (HS GRAD)**
➤ *Display:* **Column %**

| | | 90) AGE 65+ | |
		18-64	65+
146)	0-5	51.1%	37.2%
# C R C T W O R D	6	24.9%	24.5%
	7 OR MORE	24.0%	38.3%
	TOTAL	100.0%	100.0%

V=0.130*

> Click the appropriate button at the bottom of the task bar to look at the second (or "next") partial table for 6) EDUCATION.

This table includes those respondents who have only a high school degree. Again, examine the results before continuing.

➤ *View:* **Tables (SOME COLL.)**
➤ *Display:* **Column %**

| | | 90) AGE 65+ | |
		18-64	65+
146)	0-5	30.9%	21.4%
# C R C T W O R D	6	27.9%	33.9%
	7 OR MORE	41.2%	44.6%
	TOTAL	100.0%	100.0%

V=0.066

> Again, click the appropriate button at the bottom of the task bar to look at the third (or "next") partial table for 6) EDUCATION.

And this table includes those respondents having only some college. Let's look at the final table, college graduates.

➤ *View:* **Tables (COLL GRAD)**
➤ *Display:* **Column %**

| | | 90) AGE 65+ | |
		18-64	65+
146)	0-5	12.8%	15.6%
# C R C T W O R D	6	16.8%	6.3%
	7 OR MORE	70.4%	78.1%
	TOTAL	100.0%	100.0%

V=0.071

> Click the appropriate button at the bottom of the task bar to look at the last (or "next") partial table for 6) EDUCATION.

Our suspicions are confirmed. Once education is taken into account, the elderly's vocabulary scores are not lower than those of their younger counterparts. In fact, they generally tend to be higher.

Now let's turn to views on a few issues that particularly affect the elderly. Earlier we looked at international differences in support for euthanasia. The GSS has a similar question: "When a person has a disease that cannot be cured, do you think doctors should be allowed by law to end the patient's life by some painless means if the patient and his family request it?"

Data File: **GSS**
➤ Task: **Univariate**
➤ Primary Variable: **77) EUTHANASIA**
➤ View: **Pie**

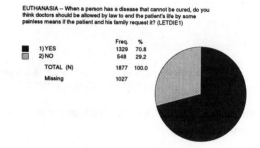

About 71 percent of the sample responds Yes to this question.

➤ Data File: **HISTORY**
➤ Task: **Historical Trends**
➤ Variable: **11) EUTHANASIA**

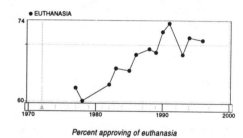

Percent approving of euthanasia

This figure has generally risen since the 1970s.

Let's see whether age affects views on euthanasia. Do you think the elderly should be more in favor or less in favor?

➤ Data File: **GSS**
➤ Task: **Cross-tabulation**
➤ Row Variable: **77) EUTHANASIA**
➤ Column Variable: **90) AGE 65+**
➤ View: **Tables**
➤ Display: **Column %**

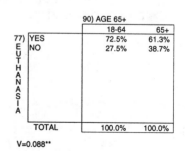

Discovering Sociology

Those over 65 are *less* likely (61.3 percent) than their younger counterparts (72.5 percent) to approve of euthanasia (V = .09**).

In the NATIONS data set we saw that religiosity, as measured by religious attendance, was negatively related to approval of euthanasia. Will we find a similar relationship in the GSS? We'll use a measure of religious attendance as our independent or column variable.

	Data File:	**GSS**
	Task:	**Cross-tabulation**
	Row Variable:	**77) EUTHANASIA**
➤	Column Variable:	**127) ATTEND**
	➤ View:	**Tables**
	➤ Display:	**Column %**

127) ATTEND

77) EUTHANASIA	NEVER	MONTH/YR	WEEKLY
YES	84.5%	79.3%	49.8%
NO	15.5%	20.7%	50.2%
TOTAL	100.0%	100.0%	100.0%

V=0.315**

The higher the religiosity, the lower the support for euthanasia (V = .32**).

As our population ages, adult children of aging parents are increasingly having to make important decisions about where their parents should live. The GSS asks, "As you know, many older people share a home with their grown children. Do you think this is generally a good idea or a bad idea?"

Do you think older people will be more likely than younger ones to think it's a good idea?

	Data File:	**GSS**
	Task:	**Cross-tabulation**
➤	Row Variable:	**67) LIVE W KID**
➤	Column Variable:	**90) AGE 65+**
	➤ View:	**Tables**
	➤ Display:	**Column %**

90) AGE 65+

67) LIVE W KID	18-64	65+
GOOD IDEA	62.1%	37.8%
BAD IDEA	37.9%	62.2%
TOTAL	100.0%	100.0%

V=0.179**

People over 65 are *less* likely than those under 65 to think that it's a good idea for older people to share a home with their grown children (V = .18**). Did you expect this result?

WORKSHEET

NAME:

COURSE:

DATE:

EXERCISE

9

REVIEW QUESTIONS

Based on the first part of this exercise, answer True or False to the following items:

The chances of living into your 70s do not generally depend on the region of the world in which you live.	T	F
The more religious a nation's people, the more they approve of euthanasia.	T	F
In the United States, people are more likely to live in a nursing home on the West Coast than in other regions of the country.	T	F
Older Americans are generally less happy than younger Americans.	T	F
Older Americans are less likely than younger Americans to approve of euthanasia.	T	F
Older Americans feel lonelier than younger Americans.	T	F

Based on the analyses in this chapter so far, would you say that the elderly have worse lives than the younger people? Why or why not?

EXPLORIT QUESTIONS

Many people think that the elderly hold more conservative views than younger people on controversial social issues having to do with morality, partly because they grew up during a time when views were more conservative than they are now. Let's explore possible age differences on several of these issues.

1.   We'll first consider sex education in the schools.

> ➤ *Data File:* **GSS**
> ➤ *Task:* **Cross-tabulation**
> ➤ *Row Variable:* **70) SEX ED?**
> ➤ *Column Variable:* **90) AGE 65+**
> ➤ *View:* **Tables**
> ➤ *Display:* **Column %**

a. What percent of older people are against sex education in the public schools? _____ %

b. What percent of younger people are against sex education in the public schools? _____ %

c. Is V statistically significant?   Yes   No

d. Do the elderly hold more conservative views on sex education?   Yes   No

2.   Now we'll examine possible age differences in views on divorce.

> Data File: **GSS**
> Task: **Cross-tabulation**
> ➤ Row Variable: **71) DIV.EASY?**
> ➤ Column Variable: **90) AGE 65+**
> ➤ View: **Tables**
> ➤ Display: **Column %**

a. What percent of older people feel that divorce should be made more difficult? _____ %

b. What percent of younger people feel that divorce should be made more difficult? _____ %

c. Is V statistically significant?   Yes   No

d. Do the elderly hold more conservative views on divorce?   Yes   No

3.   Next we'll look at views on premarital sex.

> Data File: **GSS**
> Task: **Cross-tabulation**
> ➤ Row Variable: **72) PREM.SEX**
> ➤ Column Variable: **90) AGE 65+**
> ➤ View: **Tables**
> ➤ Display: **Column %**

a. What percent of older people feel that premarital sex is always wrong? _____ %

b. What percent of younger people feel that premarital sex is always wrong? _____ %

c. Is V statistically significant?   Yes   No

d. Do the elderly hold more conservative views on premarital sex?   Yes   No

4.   Now we'll consider extramarital sex.

                    Data File: **GSS**
                         Task: **Cross-tabulation**
         ➤ Row Variable: **73) XMAR.SEX**
         ➤ Column Variable: **90) AGE 65+**
                    ➤ View: **Tables**
                 ➤ Display: **Column %**

   a.   What percent of older people feel that extramarital sex is always wrong?   _____%

   b.   What percent of younger people feel that extramarital sex is always wrong?   _____%

   c.   Is V statistically significant?   Yes   No

   d.   Do the elderly hold more conservative views on extramarital sex?   Yes   No

5.   Are there age differences in views on homosexuality?

                    Data File: **GSS**
                         Task: **Cross-tabulation**
         ➤ Row Variable: **74) HOMO.SEX**
         ➤ Column Variable: **90) AGE 65+**
                    ➤ View: **Tables**
                 ➤ Display: **Column %**

   a.   What percent of older people feel that homosexual sex is always wrong?   _____%

   b.   What percent of younger people feel that homosexual sex is always wrong?   _____%

   c.   Is V statistically significant?   Yes   No

   d.   Do the elderly hold more conservative views on homosexual sex?   Yes   No

6.   Finally, we'll examine views on the legalization of marijuana.

                    Data File: **GSS**
                         Task: **Cross-tabulation**
         ➤ Row Variable: **119) GRASS?**
         ➤ Column Variable: **90) AGE 65+**
                    ➤ View: **Tables**
                 ➤ Display: **Column %**

Exercise 9: Age and Aging   133

a. What percent of older people oppose the legalization of marijuana?   _____%

b. What percent of younger people oppose the legalization of marijuana?   _____%

c. Is V statistically significant?   Yes   No

d. Do the elderly hold more conservative views on marijuana?   Yes   No

e. Based on the analyses in these worksheets, do you think the elderly generally hold more conservative views on social and moral issues than younger people? Why or why not?

7. Now let's turn for the moment to the STATES data set.

> ➤ *Data File:* **STATES**
> ➤ *Task:* **Scatterplot**
> ➤ *Dependent Variable:* **29) %NO RELIG.**
> ➤ *Independent Variable:* **5) %>AGE 64**
> ➤ *View:* **Reg. Line**

a. What hypothesis is this scatterplot testing? (Make sure to examine the complete variable descriptions.)

b. What is the value of r?   r = _____

c. Is r statistically significant?   Yes   No

d. Is the age composition of states related to the religiosity of their residents?   Yes   No

8. In the beginning of this exercise we saw that life expectancy varies greatly around the world. What accounts for this variation? An obvious possibility is the wealth or poverty of a nation. Let's see whether the annual national product per capita, the variable we used to measure global stratification in Exercise 6, is related to life expectancy. We'll hypothesize that wealthier nations should have longer life expectancies than poorer nations.

> ➤ *Data File:* **NATIONS**
> ➤ *Task:* **Scatterplot**
> ➤ *Dependent Variable:* **19) LIFE EXPCT**
> ➤ *Independent Variable:* **29) $ PER CAP**
> ➤ *View:* **Reg. Line**

a. What is the value of r?   r = _____

b. Is this a weak, moderate, or strong relationship? (Circle one.)   Weak

                                                                                    Moderate

                                                                                    Strong

c. If wealthier nations do have longer life expectancies (or, to turn it around, if poorer nations have shorter life expectancies), what, in your opinion, are the one or two most important reasons for these differences?

9. Perhaps you just mentioned better nutrition as one reason for why wealthier nations have longer life expectancies. If so, a scatterplot should reveal a positive correlation between better nutrition and longer life expectancy. The NATIONS data set includes a measure of the daily available calories per person (based on a nation's food supply), a rough measure of the nutritional intake in each nation. Let's see whether the nations with higher caloric intake have longer life expectancies.

> Data File: **NATIONS**
> Task: **Scatterplot**
> Dependent Variable: **19) LIFE EXPCT**
> ➤ Independent Variable: **23) CALORIES**
> ➤ View: **Reg. Line**

a. In the space below, summarize the conclusion you draw from this scatterplot.

b. To find out where the United States appears in the scatterplot, click on [Find Case], then select United States from the window that appears, then click [OK]. The dot for the United States will then appear in the scatterplot. Where do you find it?   Upper left

                                                                                    Upper right

                                                                                    Lower left

                                                                                    Lower right

10. Because the elderly may lack money for transportation or have reduced physical mobility, they may find it more difficult than younger people to engage in certain leisure-time activities. Let's see whether this is true for going out to eat at a restaurant. We'll use a GSS variable that asked respondents whether they've eaten out in the last week.

> *Data File:* **GSS**
> *Task:* **Cross-tabulation**
> *Row Variable:* **158) EAT OUT**
> *Column Variable:* **90) AGE 65+**
> *View:* **Tables**
> *Display:* **Column %**

a. People aged 65 and older are less likely than those under 65 to have eaten out in the last week? (Circle one.)   T   F

b. Is the age difference here statistically significant? (Circle one.)   Yes   No

11. Several of the examples in the preliminary section and worksheets of this exercise address various beliefs about the elderly. Write a brief essay in which you use these examples to discuss the extent to which our beliefs about the elderly turn out not to be true when tested with actual data. (Be sure to provide specific statistical evidence to support your answer.)

◆ EXERCISE 10 ◆
THE FAMILY

Tasks: Mapping, Scatterplot, Univariate, Cross-tabulation, Historical Trends
Data Files: NATIONS, STATES, GSS, HISTORY

Perhaps our most important social institution is the family. Certainly it is the one with which we've had the most contact. Most of us are born into a family and are raised by it until we near the end of our teenage years. Most people later get married and have children or at least live with someone without getting married. The family provides not only food, clothing, shelter, and other necessities, but also socialization, that essential building block of society. As a primary group, the family also provides emotional support for its members.

For better or worse, however, the family has been in a state of flux. Divorce rates began rising dramatically in the 1960s and have leveled off only recently. Single-parent households, either as the result of divorce or as the result of births outside of wedlock, began to increase at about the same time. So did cohabitation, or living together in a romantic relationship without being married. All these trends have been the subject of considerable, often heated, debate that shows no signs of diminishing.

This chapter examines some key aspects of the family in contemporary life. We'll first look at international variation in some views about the family, and then examine state-by-state variation in divorce and some other family dimensions. We'll finally turn to the GSS to examine opinions on important family issues.

INTERNATIONAL VIEWS ON THE FAMILY

As you might expect, the nations of the world differ widely in their views on various aspects of marriage and the family. Let's first examine the percent who agree that "marriage is an outdated institution."

➤ *Data File:* **NATIONS**
➤ *Task:* **Mapping**
➤ *Variable 1:* **81) WED PASSE'**
➤ *View:* **Map**

WED PASSE' -- PERCENT WHO AGREE THAT "MARRIAGE IS AN OUTDATED INSTITUTION." (WVS)

The countries of Western Europe appear to be the most likely to agree that marriage is outdated.

Data File:	**NATIONS**
Task:	**Mapping**
Variable 1:	**81) WED PASSE'**
➤ View:	**List: Rank**

RANK	CASE NAME	VALUE
1	France	29.00
2	Brazil	27.00
3	Belgium	23.00
4	Portugal	22.00
5	Netherlands	21.00
6	United Kingdom	18.00
6	Slovenia	18.00
6	Denmark	18.00
9	Mexico	17.00
10	Belarus	16.00

Although well under half of all the nations' residents feel marriage is outdated, the French, with 29 percent, are most likely to feel this way. The United States ranks near the bottom; only 8 percent of its residents think marriage is outdated. India, with 5 percent, is least likely to view marriage as outdated. What would explain these differences?

Many of today's controversial issues about the family concern sex and sexuality. One variable in the NATIONS data set measures the percent who think it is "never" acceptable for married people to have an affair.

Data File:	**NATIONS**
Task:	**Mapping**
➤ Variable 1:	**109) EX-MARITAL**
➤ View:	**Map**

EX-MARITAL -- PERCENT WHO THINK IT IS NEVER OK FOR MARRIED PEOPLE TO HAVE AN AFFAIR (WVS)

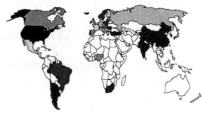

Generally, the European nations are least likely to feel this way, but even in Europe there appears to be much variation.

Data File:	**NATIONS**
Task:	**Mapping**
Variable 1:	**109) EX-MARITAL**
➤ View:	**List: Rank**

RANK	CASE NAME	VALUE
1	India	91.00
2	Turkey	87.00
3	China	72.00
4	South Korea	71.00
4	Iceland	71.00
6	South Africa	70.00
6	United States	70.00
6	Argentina	70.00
9	Ireland	69.00
10	Poland	67.00

There's quite a range of views on adultery. In India, 91 percent of the public feels that adultery is never acceptable, with the United States slightly behind at 70 percent. In Canada, just more than half, 53 percent, feel this way. At the other end, only 26 percent of the Czech Republic feels adultery is never acceptable.

Another controversial sexual issue today is homosexuality. The NATIONS data set includes a measure of the percent who believe homosexuality is "never" acceptable behavior.

Data File: **NATIONS**
Task: **Mapping**
➤ Variable 1: **111) GAY SEX**
➤ View: **Map**

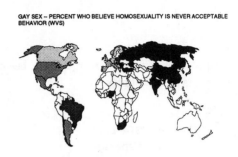

GAY SEX – PERCENT WHO BELIEVE HOMOSEXUALITY IS NEVER ACCEPTABLE BEHAVIOR (WVS)

Disapproval of homosexuality is lowest in Canada and the nations of Western Europe.

One reason for this international variation in views about homosexuality might be differences in education. Perhaps as people become more educated, they become less disapproving of homosexuality. If so, the most educated nations would be less likely to say homosexuality is never acceptable. Let's see whether this is so.

Data File: **NATIONS**
➤ Task: **Scatterplot**
➤ Dependent Variable: **111) GAY SEX**
➤ Independent Variable: **40) EDUCATION**
➤ View: **Reg. Line**

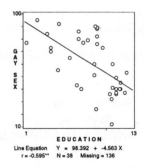

EDUCATION
Line Equation   Y = 98.392 + -4.563 X
r = -0.595**   N = 38   Missing = 136

Our hypothesis is supported: the more educated a nation's residents, the less likely they are to say that homosexuality is never acceptable (r = –.60**).

MARRIAGE AND THE FAMILY IN THE UNITED STATES

Let's take a look at which states have the highest marriage rates. We'll use a measure of marriages per 1000 population.

> *Data File:* **STATES**
> *Task:* **Mapping**
> *Variable 1:* **35) MARRIAGES**
> *View:* **Map**

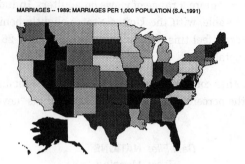

MARRIAGES -- 1989: MARRIAGES PER 1,000 POPULATION (S.A.,1991)

No clear geographic pattern emerges here, because the states with the highest marriage rates are in different parts of the country.

Data File: **STATES**
Task: **Mapping**
Variable 1: **35) MARRIAGES**
> *View:* **List: Rank**

RANK	CASE NAME	VALUE
1	Nevada	106.3
2	Hawaii	16.2
3	South Carolina	15.5
4	Arkansas	14.4
5	Kentucky	13.5
6	Tennessee	13.2
7	Idaho	12.9
8	Virginia	11.3
9	Arizona	11.2
10	Alaska	11.0

Nevada's marriage rate soars far above any other state. The popular image of people getting married abruptly in Reno might not be too far off the mark!

We can also see which states have the highest proportion of divorced people.

Data File: **STATES**
Task: **Mapping**
> *Variable 1:* **38) %DIVORCED**
> *View:* **Map**

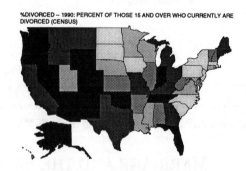

%DIVORCED -- 1990: PERCENT OF THOSE 15 AND OVER WHO CURRENTLY ARE DIVORCED (CENSUS)

Divorce is most common west of the Mississippi. Why do you think that's so? Is there something about living out west that helps lead to divorce? Do divorced people tend to move west? Or is something else going on?

Discovering Sociology

We turn now to the GSS and begin by continuing to look at divorce. We'll first see what percent of the sample has ever been divorced or separated. (Note that people who have *never* been married are excluded from this calculation.)

➤ *Data File:* **GSS**
➤ *Task:* **Univariate**
➤ *Primary Variable:* **91) DIVORCED**
➤ *View:* **Pie**

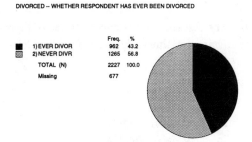

DIVORCED -- WHETHER RESPONDENT HAS EVER BEEN DIVORCED

	Freq.	%
■ 1) EVER DIVOR	962	43.2
▨ 2) NEVER DIVR	1265	56.8
TOTAL (N)	2227	100.0
Missing	677	

About 43 percent of those who have ever been married have been divorced/separated (hereafter referred to as just "divorced").

What predicts your chances of getting divorced? We often hear that your chances are greater if your own parents were divorced. Let's see whether GSS data support this hypothesis.

Data File: **GSS**
➤ *Task:* **Cross-tabulation**
➤ *Row Variable:* **91) DIVORCED**
➤ *Column Variable:* **92) PARS.DIV?**
➤ *View:* **Tables**
➤ *Display:* **Column %**

92) PARS. DIV?

91) DIVORCED	TOGETHER	DIVORCED
EVER DIVOR	40.4%	52.7%
NEVER DIVR	59.6%	47.3%
TOTAL	100.0%	100.0%

V=0.088**

People whose parents were divorced are 12 percentage points more likely to get divorced themselves (V = .09**).

Earlier we saw that states with nonreligious residents had higher divorce rates. Let's see whether the less religious individuals in the GSS are more likely to be divorced.

Data File: **GSS**
Task: **Cross-tabulation**
Row Variable: **91) DIVORCED**
➤ *Column Variable:* **127) ATTEND**
➤ *View:* **Tables**
➤ *Display:* **Column %**

127) ATTEND

91) DIVORCED	NEVER	MONTH/YR	WEEKLY
EVER DIVOR	53.5%	45.9%	34.8%
NEVER DIVR	46.5%	54.1%	65.2%
TOTAL	100.0%	100.0%	100.0%

V=0.131**

People who never go to religious services are more likely (53.5 percent) than those who often attend religious services (34.8 percent) to be divorced (V = .13**). Note that we can't necessarily conclude that religiosity affects divorce, because people might become less religious after getting divorced. The relationship might also be spurious if some third factor, say income, affects both religiosity and the chances for divorce.

Now let's just compare people in terms of their *current* marital status. Which group do you think will be the happiest: those who are married, widowed, divorced (or separated), or single? Let's find out.

    Data File: **GSS**
    Task: **Cross-tabulation**
➤ *Row Variable:* **132) HAPPY?**
➤ *Column Variable:* **54) MARITAL**
    ➤ *View:* **Tables**
    ➤ *Display:* **Column %**

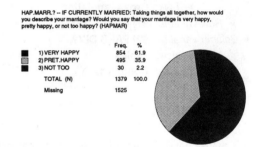

54) MARITAL				
132)	MARRIED	WIDOWED	DIV/SEP.	NEV. MARR
VERY HAPPY	41.5%	25.2%	17.0%	20.8%
PRET.HAPPY	52.5%	53.6%	63.9%	64.1%
NOT TOO	5.9%	21.2%	19.1%	15.1%
TOTAL	100.0%	100.0%	100.0%	100.0%

V=0.195**

Marital status is certainly related to happiness (V = .20**); married people are the most likely to say they're very happy.

The GSS asks married people, "Taking things all together, how would you describe your marriage? Would you say that your marriage is very happy, pretty happy, or not too happy?"

    Data File: **GSS**
    ➤ *Task:* **Univariate**
➤ *Primary Variable:* **63) HAP.MARR.?**
    ➤ *View:* **Pie**

HAP.MARR.? -- IF CURRENTLY MARRIED: Taking things all together, how would you describe your marriage? Would you say that your marriage is very happy, pretty happy, or not too happy? (HAPMAR)

		Freq.	%
■	1) VERY HAPPY	854	61.9
▨	2) PRET.HAPPY	495	35.9
■	3) NOT TOO	30	2.2
	TOTAL (N)	1379	100.0
	Missing	1525	

About 62 percent of married people describe their marriage as very happy.

➤ *Data File:* **HISTORY**
    ➤ *Task:* **Historical Trends**
➤ *Variable:* **13) HAP.MARR**

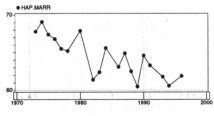

Percent saying their marriage is very happy

This percent has declined somewhat since a quarter century ago.

Does family income predict marital happiness?

> *Data File:* **GSS**
> > *Task:* **Cross-tabulation**
> *Row Variable:* **63) HAP.MARR.?**
> *Column Variable:* **7) FAM INCOME**
> > *View:* **Tables**
> > *Display:* **Column %**

	7) FAM INCOME		
	$1K-$19999	$20K-39999	$40K & UP
63) VERY HAPPY	64.1%	63.5%	59.8%
HAP PRET.HAPPY	34.0%	34.8%	37.6%
P NOT TOO	2.0%	1.7%	2.6%
MARR?			
TOTAL	100.0%	100.0%	100.0%

V=0.032

The social class differences here are not statistically significant (V = .03).

Do you think religious people will be more likely to report happier marriages?

Data File: **GSS**
Task: **Cross-tabulation**
Row Variable: **63) HAP.MARR.?**
> *Column Variable:* **127) ATTEND**
> *View:* **Tables**
> *Display:* **Column %**

	127) ATTEND		
	NEVER	MONTH/YR	WEEKLY
63) VERY HAPPY	55.8%	58.9%	69.1%
H PRET.HAPPY	41.8%	38.5%	29.6%
A P NOT TOO	2.4%	2.7%	1.3%
MARR?			
TOTAL	100.0%	100.0%	100.0%

V=0.078**

People who attend religious services weekly are more likely (69.1 percent) to say their marriages are very happy than those who never attend (55.8 percent). The results are statistically significant (V = .08**).

In the NATIONS data set, we examined attitudes on adultery. Let's see what respondents say in the GSS.

Data File: **GSS**
> *Task:* **Univariate**
> *Primary Variable:* **73) XMAR.SEX**
> *View:* **Pie**

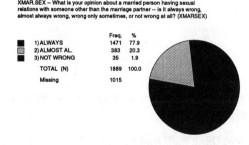

XMAR.SEX -- What is your opinion about a married person having sexual relations with someone other than the marriage partner -- is it always wrong, almost always wrong, wrong only sometimes, or not wrong at all? (XMARSEX)

	Freq.	%
1) ALWAYS	1471	77.9
2) ALMOST AL.	383	20.3
3) NOT WRONG	35	1.9
TOTAL (N)	1889	100.0
Missing	1015	

Almost 78 percent of the U.S. public thinks adultery is always wrong.

> *Data File:* **HISTORY**
>> *Task:* **Historical Trends**
> *Variable:* **14) XMAR.SEX**

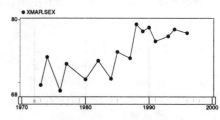

Percent saying extramarital sex is always wrong

This percent has generally risen during the last quarter century.

Now let's look at views on premarital sex. Traditionally, sex was considered acceptable only if it occurred between spouses. Many people still feel that way, but many others think sex outside of marriage is acceptable. The GSS asked what respondents think about a man and woman having "sex relations before marriage."

> *Data File:* **GSS**
>> *Task:* **Univariate**
> *Primary Variable:* **72) PREM.SEX**
>> *View:* **Pie**

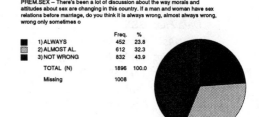

Almost 24 percent think premarital sex is always wrong, about 32 percent think it's almost always wrong, and some 44 percent think it's not wrong.

> *Data File:* **HISTORY**
>> *Task:* **Historical Trends**
> *Variable:* **15) PREM.SEX**

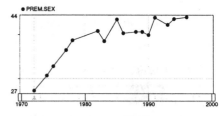

Percent saying premarital sex is not wrong

The figure for thinking that premarital sex is *not* wrong has risen by about 17 percentage points in the last quarter century.

Do you think we'll find a gender difference on views about premarital sex?

Discovering Sociology

➤ *Data File:* **GSS**
➤ *Task:* **Cross-tabulation**
➤ *Row Variable:* **72) PREM.SEX**
➤ *Column Variable:* **45) GENDER**
➤ *View:* **Tables**
➤ *Display:* **Column %**

		45) GENDER	
		FEMALE	MALE
72) PREM.SEX	ALWAYS	26.8%	20.1%
	ALMOST AL.	33.8%	30.3%
	NOT WRONG	39.4%	49.6%
	TOTAL	100.0%	100.0%

V=0.107**

Women are more likely than men to oppose premarital sex (V = .11**). Why do you think this difference exists?

What about religiosity? Do you think the more religious people are, the more they should think premarital sex is wrong?

Data File: **GSS**
Task: **Cross-tabulation**
Row Variable: **72) PREM.SEX**
➤ *Column Variable:* **127) ATTEND**
➤ *View:* **Tables**
➤ *Display:* **Column %**

		127) ATTEND		
		NEVER	MONTH/YR	WEEKLY
72) PREM.SEX	ALWAYS	10.0%	13.8%	47.5%
	ALMOST AL.	23.0%	35.5%	32.1%
	NOT WRONG	67.0%	50.7%	20.5%
	TOTAL	100.0%	100.0%	100.0%

V=0.299**

Religiosity is strongly related to views on premarital sex, because people who attend religious services weekly are more likely (47.5 percent) than those who never attend (10.0 percent) to think premarital sex is always wrong (V = .30**).

EDUCATION AND RELIGION

Tasks: Mapping, Scatterplot, Cross-tabulation
Data Files: NATIONS, STATES, GSS

As social institutions, education and religion play a very important role in most people's lives. First, they both function as agents of socialization. Through schooling and religious training we learn many of our values and beliefs. As we've seen in some previous chapters, education is often associated with acceptance of nontraditional beliefs and practices, whereas religiosity is often associated with just the opposite.

Second, both education and religion help to integrate our society. Not only do we learn common values from both institutions, we also interact with others as we go to school and attend religious services. As Emile Durkheim emphasized long ago, such social interaction is important for social stability.

Third, both education and religion help determine our life chances. More educated people typically have greater opportunities for high-paying jobs and the benefits that high incomes often bring. Unfortunately, the opposite is true as well: lack of education often dooms people to low incomes, poor health, and other negative life chances. In the United States and elsewhere, religious affiliation has historically been associated with one's placement on the socioeconomic ladder. In the United States, for example, certain Protestant denominations have enjoyed high socioeconomic status and political power, while other Protestant denominations have ranked lower on both scores. Although times have changed, historically U.S. Catholics and Jews were victims of discrimination in the schools and the workplace and also victims of violence directed at them because of their religion. Hate crimes against Jews continue to occur.

Education and religion around the world exhibit similar characteristics. The level of a nation's education is critical for its people's life chances and often for their beliefs. Religious affiliation and practice have a similar international impact. To understand the world today, it's important to appreciate the difference that education and religion make.

This exercise explores some key aspects of education and religion across the world and within the United States. We will primarily focus on what difference they make for life chances and beliefs. We'll first look at education and then at religion and, within each institution's discussion, follow our usual practice of starting with the NATIONS data set and then turning to our two U.S. data sets.

EDUCATION

International Differences in Education and Its Correlates

Let's first map education to see which parts of the world are most and least educated.

➤ Data File: **NATIONS**
➤ Task: **Mapping**
➤ Variable 1: **40) EDUCATION**
➤ View: **Map**

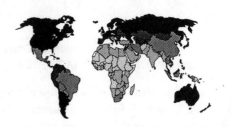

EDUCATION -- 1990: MEAN YEARS OF SCHOOL AMONG 25 AND OLDER (HDR, 1993)

The countries of North America and Western Europe clearly have the highest average education and Africa generally has the lowest.

As noted above, we've already seen that international differences in education are often associated with differences in acceptance of nontraditional beliefs and practices. Let's see whether education is associated with one additional belief, abortion. The NATIONS data set includes a measure of the percent who approve of an abortion when the mother's health is at risk.

Data File: **NATIONS**
➤ Task: **Scatterplot**
➤ Dependent Variable: **16) MOM HEALTH**
➤ Independent Variable: **40) EDUCATION**
➤ View: **Reg. Line**

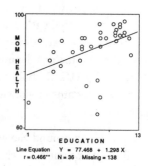

Line Equation   Y = 77.468 + 1.298 X
r = 0.466**   N = 36   Missing = 138

The more educated a nation's residents are, the more likely they are to approve of abortion when the mother's health is in danger (r = .47**).

Education in the United States

Has it ever occurred to you that people in some regions of the United States are more educated than those in other regions? Let's take a look. Our first map will be the percent of a state's 25-and-over population who have at least a college degree, and the second map will be the percent who never completed high school.

➤ *Data File:* **STATES**
➤ *Task:* **Mapping**
➤ *Variable 1:* **75) COLL.DEGR**
➤ *View:* **Map**

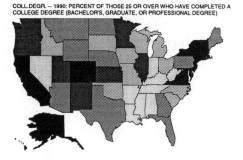

COLL.DEGR. -- 1990: PERCENT OF THOSE 25 OR OVER WHO HAVE COMPLETED A
COLLEGE DEGREE (BACHELOR'S, GRADUATE, OR PROFESSIONAL DEGREE)

College graduates are most likely to be found in California, Colorado, and several states in the East.

Data File: **STATES**
Task: **Mapping**
➤ *Variable 1:* **76) DROPOUTS**
➤ *View:* **Map**

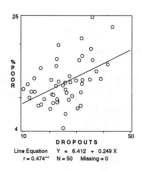

DROPOUTS -- 1990: PERCENT OF PERSONS WHO LEFT SCHOOL WITHOUT
GRADUATING FROM HIGH SCHOOL (WA, 1993)

The southern states have the highest proportion of high school dropouts. Let's see if the least educated states also the poorest?

Data File: **STATES**
➤ *Task:* **Scatterplot**
➤ *Dependent Variable:* **55) %POOR**
➤ *Independent Variable:* **76) DROPOUTS**
➤ *View:* **Reg. Line**

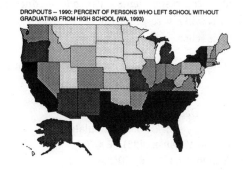

Line Equation   Y = 6.412 + 0.249 X
r = 0.474**   N = 50   Missing = 0

The higher the proportion of high school dropouts, the poorer the state (r = .47**). Note that we can't necessarily conclude that low education produces high poverty, because high poverty may lead to lower education.

Let's turn now to the GSS. We'll first see whether your parents' education influences the education you eventually acquire. Our independent variable has three categories: both parents lacked a high school degree, both parents had a high school degree, and both parents had at least a college degree. For simplicity's sake, respondents whose parents had different degrees are excluded.

➤ *Data File:* **GSS**
➤ *Task:* **Cross-tabulation**
➤ *Row Variable:* **6) EDUCATION**
➤ *Column Variable:* **20) PARSDEGREE**
➤ *View:* **Tables**
➤ *Display:* **Column %**

20) PARSDEGREE

		BOTH <HS	BOTH HS	BOTH BA/G
6)	NO HS GRAD	32.0%	7.0%	3.2%
E D U C A T I O N	HS GRAD	38.7%	30.4%	5.3%
	SOME COLL.	15.6%	31.5%	21.3%
	COLL GRAD	13.7%	31.1%	70.2%
	TOTAL	100.0%	100.0%	100.0%

V=0.338**

Your parents' education is very strongly related to your own education (V = .34**). While only 13.7 percent of the respondents with parents who dropped out of high school attained college degrees, 70.2 percent of the respondents with college-educated parents did so. Based on these figures, we are five times more likely to get a college degree if our parents went to college than if they didn't graduate from high school.

Education is important for many aspects of your life. Let's see if it's associated with health.

Data File: **GSS**
Task: **Cross-tabulation**
➤ *Row Variable:* **133) HEALTH**
➤ *Column Variable:* **6) EDUCATION**
➤ *View:* **Tables**
➤ *Display:* **Column %**

6) EDUCATION

		NO HS GRA	HS GRAD	SOME COL	COLL GRA
133)	EXCELLENT	15.3%	27.4%	30.9%	45.3%
H E A L T H	GOOD	42.5%	50.3%	55.7%	45.6%
	FAIR/POOR	42.2%	22.3%	13.5%	9.2%
	TOTAL	100.0%	100.0%	100.0%	100.0%

V=0.226**

Education is very much related to people's health, because those with a college degree are more likely (45.3 percent) than those without a high school degree (15.3 percent) to report excellent health (V = .23**). Why do you think this relationship is so strong?

A traditional belief persisting among some segments of our population is that a woman's place is in the home. Let's see whether education is related to agreement with the statement that "women should take care of their homes and leave running the country to men."

Data File: **GSS**
Task: **Cross-tabulation**
➤ *Row Variable:* **46) WOMEN HOME**
➤ *Column Variable:* **6) EDUCATION**
➤ *View:* **Tables**
➤ *Display:* **Column %**

6) EDUCATION

		NO HS GRA	HS GRAD	SOME COL	COLL GRA
46)	AGREE	33.0%	19.0%	13.3%	6.5%
W O M E N	DISAGREE	67.0%	81.0%	86.7%	93.5%
H O M E	TOTAL	100.0%	100.0%	100.0%	100.0%

V=0.237**

People with a college degree are less likely (6.5 percent) than those without a high school degree (33.0 percent) to agree that a woman's place is in the home (V = .24**).

RELIGION

Religion and Religious Belief in International Perspective

The NATIONS data set includes several variables dealing with religion. Let's map the distribution of two religions, Christianity and Islam.

➤ *Data File:* **NATIONS**
   ➤ *Task:* **Mapping**
➤ *Variable 1:* **43) %CHRISTIAN**
     ➤ *View:* **Map**

%CHRISTIAN -- PERCENT CHRISTIAN (WCE)

This map shows the percent of a nation's population that is Protestant or Catholic. Christianity is obviously spread throughout the world, but is most common in the Western Hemisphere and parts of Europe.

   Data File: **NATIONS**
     Task: **Mapping**
➤ *Variable 1:* **42) %MUSLIM**
     ➤ *View:* **Map**

%MUSLIM -- PERCENT MUSLIM (WCE)

This map is pretty much the opposite of the one we just saw. The Muslim nations tend to be in northern Africa, the Middle East, and parts of Asia.

Earlier we saw that the more educated nations were more approving of abortion. Let's see what difference religiosity makes. Do you think we'll find the opposite result—that the more religious nations will be less approving of abortion? Our measure of religiosity will be the percent who say that God is important in their lives.

Data File: **NATIONS**

Task: **Scatterplot**

➤ Dependent Variable: **16) MOM HEALTH**

➤ Independent Variable: **52) GOD IMPORT**

➤ View: **Reg. Line**

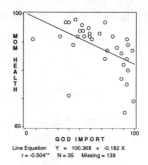

Line Equation   Y = 100.368 + -0.182 X
r = -0.504**   N = 35   Missing = 139

Our hypothesis is supported: the more religious a nation, the less it approves of an abortion when the mother's health is in danger (r = −.50**).

Religion in the United States

Your STATES data set contains several variables of interest for an understanding of religion and religiosity in the United States. We'll look at just a couple.

➤ Data File: **STATES**

➤ Task: **Mapping**

➤ Variable 1: **33) CH.MEMBERS**

➤ View: **Map**

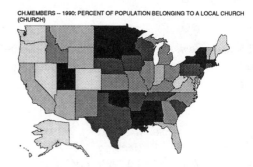

CH.MEMBERS -- 1990: PERCENT OF POPULATION BELONGING TO A LOCAL CHURCH (CHURCH)

Church membership is lowest in the western part of the nation, with the notable exception of Utah.

Data File: **STATES**

Task: **Mapping**

Variable 1: **33) CH.MEMBERS**

➤ View: **List: Rank**

RANK	CASE NAME	VALUE
1	Utah	79.8
2	Rhode Island	76.7
3	North Dakota	75.9
4	Alabama	71.0
5	Louisiana	70.5
6	Mississippi	70.2
7	South Dakota	68.1
8	Oklahoma	66.8
9	New York	65.7
10	Massachusetts	65.5

Utah and Rhode Island lead the nation in church membership; almost 80 percent of their residents belong to a local church. At the other end lie Alaska, Nevada, and Oregon, where only 32 percent belong. How would you explain these regional differences?

Let's turn now to the GSS. We'll first examine a few ways in which the United States' three major religions differ from one another, and then compare Protestant denominations.

Let's see whether Protestants, Catholics, and Jews differ in their educational levels.

➤ *Data File:* **GSS**
➤ *Task:* **Cross-tabulation**
➤ *Row Variable:* **6) EDUCATION**
➤ *Column Variable:* **126) RELIGION**
➤ *View:* **Tables**
➤ *Display:* **Column %**

126) RELIGION

6) EDUCATION	PROTESTA	CATHOLIC	JEWISH
NO HS GRAD	19.2%	14.5%	5.9%
HS GRAD	30.9%	28.8%	13.2%
SOME COLL.	27.3%	29.0%	16.2%
COLL GRAD	22.6%	27.7%	64.7%
TOTAL	100.0%	100.0%	100.0%

V=0.121**

Catholics have slightly higher levels of education than Protestants, and Jews have higher levels of education than the other two groups (V = .12**).

Do you think Catholics should be more likely than the other two groups to oppose abortion? The GSS asks respondents whether they think a woman should be allowed to have a legal abortion if "she wants it for any reason."

Data File: **GSS**
Task: **Cross-tabulation**
➤ *Row Variable:* **68) ABORT ANY**
➤ *Column Variable:* **126) RELIGION**
➤ *View:* **Tables**
➤ *Display:* **Column %**

126) RELIGION

68) ABORT ANY	PROTESTA	CATHOLIC	JEWISH
YES	40.4%	42.1%	86.5%
NO	59.6%	57.9%	13.5%
TOTAL	100.0%	100.0%	100.0%

V=0.143**

Catholics do oppose abortion more than Jews, but so do Protestants (V = .14**); Catholics are not more likely than Protestants to oppose it.

Now let's focus on religiosity. Will the more religious people be more likely to oppose abortion? To find out, we'll use a measure of the frequency of praying.

Data File: **GSS**
Task: **Cross-tabulation**
Row Variable: **68) ABORT ANY**
➤ Column Variable: **128) PRAY**
➤ View: **Tables**
➤ Display: **Column %**

128) PRAY			
68) ABORT ANY		DAILY	WEEKLY	< WEEKLY
YES		36.0%	55.0%	63.4%
NO		64.0%	45.0%	36.6%
TOTAL		100.0%	100.0%	100.0%

V=0.235**

Religiosity is strongly related to abortion opposition, because those who pray daily are more likely (64.0 percent) than those who pray less than weekly (36.6 percent) to oppose a legal abortion for a woman who wants one for any reason (V = .24**).

Is education related to religious beliefs about the Bible? The GSS asks respondents to choose one of the following: (a) the Bible is the actual word of God and is to be taken literally; (b) the Bible is the inspired word of God but not everything in it should be taken literally; and (c) the Bible is an ancient book of fables . . . recorded by men.

Data File: **GSS**
Task: **Cross-tabulation**
➤ Row Variable: **130) BIBLE**
➤ Column Variable: **6) EDUCATION**
➤ View: **Tables**
➤ Display: **Column %**

6) EDUCATION				
130) BIBLE		NO HS GRA	HS GRAD	SOME COL	COLL GRA
ACTUAL		51.3%	38.1%	26.9%	15.6%
INSPIRED		35.4%	47.1%	58.8%	57.7%
ANCI.BOOK		13.4%	14.8%	14.2%	26.7%
TOTAL		100.0%	100.0%	100.0%	100.0%

V=0.200**

Education is strongly related to belief in the Bible as the actual word of God (V = .20**). People without a high school education are over three times more likely than those with a college degree to believe the Bible is God's actual word.

◆ EXERCISE 12 ◆

WORK AND THE ECONOMY

Tasks: Mapping, Scatterplot, Univariate, Cross-tabulation
Data Files: NATIONS, STATES, GSS

Every society has an *economy*, or a system for the production and distribution of goods and services. The type of economy a society enjoys has important implications for the lives of its members. In today's world, agricultural societies are much poorer than industrial ones and, as a result, have lower life expectancies, higher rates of disease, and other such problems. A society's economy also affects its members' views on many issues, including those not related to work or the economy.

In the United States, sociologists, economists, and other scholars have carried out much research on work and employment. Many studies have examined the sources and consequences of unemployment, the determinants of job satisfaction, the aspects of work that Americans think are most important, and other matters. This body of research has yielded a rich understanding of work in U.S. society.

This exercise, then, examines work and economy in the United States and around the world. We begin by looking internationally at agricultural economies and at their implications for the lives of individuals. We then turn to the United States to explore such issues as unemployment and job satisfaction.

INTERNATIONAL ECONOMIES AND LIFE CHANGES

As noted above, many of today's nations continue to be primarily agricultural. Let's map the percent of gross domestic product (GDP) that is accounted for by agriculture to see where the agricultural nations lie.

> *Data File:* **NATIONS**
> ➤ *Task:* **Mapping**
> ➤ *Variable 1:* **32) % AGRIC $**
> ➤ *View:* **Map**

% AGRIC $ -- PERCENT OF GDP [GROSS DOMESTIC PRODUCT] ACCOUNTED FOR BY AGRICULTURE (TWF 1994)

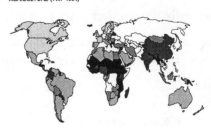

The most agriculturally dependent nations lie in Africa and parts of Asia, the least in North America and Europe.

RANK	CASE NAME	VALUE
1	Solomon Islands	70.00
2	Nepal	60.00
2	Burundi	60.00
2	Laos	60.00
5	Tanzania	58.00
6	Uganda	57.00
7	Guinea-Bissau	50.00
7	Mauritania	50.00
7	Ghana	50.00
7	Mali	50.00

Data File: **NATIONS**
Task: **Mapping**
Variable 1: **32) % AGRIC $**
➤ View: **List: Rank**

In terms of GDP, the most agricultural nation is the Solomon Islands, where 70 percent of the GDP is accounted for by agriculture. The least agricultural nation is Sweden, where only 1.2 percent of the GDP is agriculturally based. The United States ranks just above at 2.0 percent.

Another variable for mapping the agricultural nations is the percent of the labor force employed by agriculture.

Data File: **NATIONS**
Task: **Mapping**
➤ Variable 1: **33) % IN AGR.**
➤ View: **Map**

% IN AGR. -- PERCENT OF LABOR FORCE EMPLOYED BY AGRICULTURE (TWF 1994)

Not surprisingly, this map looks very similar to the previous one.

If the poorest nations in the world are agricultural, the agricultural nations should be the worst off in terms of life expectancy, illness, and other problems. Let's find out.

Data File: **NATIONS**
➤ Task: **Scatterplot**
➤ Dependent Variable: **19) LIFE EXPCT**
➤ Independent Variable: **33) % IN AGR.**
➤ View: **Reg. Line**

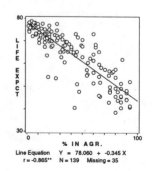

Line Equation Y = 78.060 + -0.345 X
r = -0.865** N = 139 Missing = 35

Discovering Sociology

This is one of the strongest correlations we've seen in this workbook so far (r = −.87**). The most agricultural nations have much lower life expectancies than the least agricultural ones. The latter, of course, are the world's industrial societies.

What about infant mortality?

Data File: **NATIONS**
Task: **Scatterplot**
➤ Dependent Variable: **11) INF.MORTL**
➤ Independent Variable: **33) % IN AGR.**
➤ View: **Reg. Line**

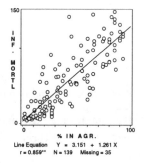

The agricultural nations have much higher levels of infant mortality (r = .86**).

Data File: **NATIONS**
Task: **Scatterplot**
➤ Dependent Variable: **40) EDUCATION**
➤ Independent Variable: **33) % IN AGR.**
➤ View: **Reg. Line**

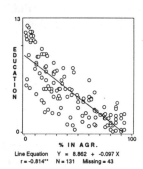

The agricultural nations have much lower levels of education (r = −.81**).

WORK AND EMPLOYMENT IN THE UNITED STATES

Do you know which states are the most agricultural? Let's examine this and other issues using the STATES data file.

> ➤ *Data File:* **STATES**
> ➤ *Task:* **Mapping**
> ➤ *Variable 1:* **70) %AGRI.EMP**
> ➤ *View:* **Map**

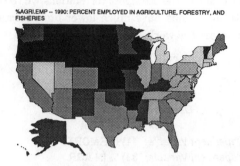

%AGRI.EMP -- 1990: PERCENT EMPLOYED IN AGRICULTURE, FORESTRY, AND FISHERIES

The agricultural, forestry, and fishery states tend to be in the upper Midwest.

At the international level, the agricultural nations have less educated populations. In the United States, do the more agricultural states have lower percentages of college graduates?

> *Data File:* **STATES**
> ➤ *Task:* **Scatterplot**
> ➤ *Dependent Variable:* **75) COLL.DEGR.**
> ➤ *Independent Variable:* **70) %AGRI.EMP**
> ➤ *View:* **Reg. Line**

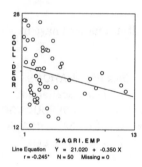

Line Equation   Y = 21.020 + -0.350 X
r = -0.245*   N = 50   Missing = 0

The relationship is relatively small but statistically significant (r = −.25*); fewer college-educated people live in the more agricultural states.

One of the toughest things about work in the United States is getting to your job in the morning and getting home at night. Many people take public transportation—a bus, train, cab, subway, and so on. Let's see which states have the highest use of public transportation.

> *Data File:* **STATES**
> ➤ *Task:* **Mapping**
> ➤ *Variable 1:* **68) PUB.TRNS**
> ➤ *View:* **Map**

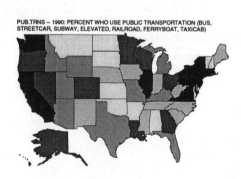

PUB.TRNS -- 1990: PERCENT WHO USE PUBLIC TRANSPORTATION (BUS, STREETCAR, SUBWAY, ELEVATED, RAILROAD, FERRYBOAT, TAXICAB)

ransportation is most often used in Washington, California, Illinois, and several eastern states.

Discovering Sociology

What accounts for this range in public transportation use? The obvious answer is urbanism: in rural areas, people are more likely to drive to work in their cars or pickups, and public transportation often doesn't exist in the first place. Let's see whether the more urban states have higher public transportation use.

Data File: **STATES**
➤ Task: **Scatterplot**
➤ Dependent Variable: **68) PUB. TRNS**
➤ Independent Variable: **11) %URBAN**
➤ View: **Reg. Line**

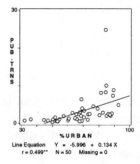

Bingo! The correlation, r, is .50**.

If you take public transportation to work, does that take more time or less time than if you drive to work or get there some other way (e.g., walking or bicycling)? Our dependent variable will be the average time it takes to get to work.

Data File: **STATES**
Task: **Scatterplot**
➤ Dependent Variable: **69) AVG. TRAVL**
➤ Independent Variable: **68) PUB. TRNS**
➤ View: **Reg. Line**

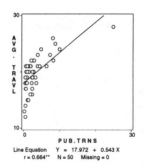

The states with the highest public transportation use tend to be the states where it takes the most time to get to work (r = .66**). Note, however, that this does not *prove* public transportation takes more time than driving. If you lived in New York City or another big city, public transportation could take less time than driving would. All the scatterplot suggests is that those who do take public transportation spend more time commuting than those who drive or use some other means to get to work. What makes the most sense for any one individual obviously depends partly on where that person lives. There's also the question of which type of commuting would cause the least pollution—but that's another question altogether.

Time now for the GSS. One of the most important aspects of a job is how much you like it. The GSS asks whether respondents are satisfied with the work they do.

➤ *Data File:* **GSS**
➤ *Task:* **Univariate**
➤ *Primary Variable:* **10) LIKE JOB?**
➤ *View:* **Pie**

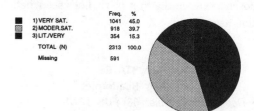

LIKE JOB? -- On the whole, how satisfied are you with the work you do -- would you say you are very satisfied, moderately satisfied, a little dissatisfied, or very dissatisfied? (SATJOB)

		Freq.	%
■	1) VERY SAT.	1041	45.0
▨	2) MODER.SAT.	918	39.7
■	3) LIT./VERY	354	15.3
	TOTAL (N)	2313	100.0
	Missing	591	

Forty-five percent of the respondents are very satisfied with their jobs, 40 percent moderately satisfied, and 15 percent dissatisfied. What affects job satisfaction? At least since the time of Karl Marx, social scientists have observed that blue-collar jobs are less creative and more alienating than white-collar jobs. If so, people working in the latter should be more satisfied with their jobs than people working in the former. Let's see if this is so, using the GSS measure of occupational prestige that roughly corresponds to the white collar–blue collar distinction.

Data File: **GSS**
➤ *Task:* **Cross-tabulation**
➤ *Row Variable:* **10) LIKE JOB?**
➤ *Column Variable:* **3) PRESTIGE**
➤ *View:* **Tables**
➤ *Display:* **Column %**

10) LIKE JOB?	3) PRESTIGE	
	< 40	40 AND UP
VERY SAT.	38.3%	50.4%
MODER.SAT.	42.4%	37.7%
LIT./VERY	19.3%	11.9%
TOTAL	100.0%	100.0%

V=0.135**

People in high-prestige jobs are more likely (50.4 percent) than those in low-prestige ones (38.3 percent) to be very satisfied with their jobs (V = .14**).

Compared to men, women are more likely to be in low-paying jobs in the service sector and are more likely to be sexually harassed in the workplace. Does that mean they should be less satisfied than men with their jobs?

Data File: **GSS**
Task: **Cross-tabulation**
Row Variable: **10) LIKE JOB?**
➤ *Column Variable:* **45) GENDER**
➤ *View:* **Tables**
➤ *Display:* **Column %**

10) LIKE JOB?	45) GENDER	
	FEMALE	MALE
VERY SAT.	44.4%	45.7%
MODER.SAT.	39.4%	40.0%
LIT./VERY	16.2%	14.3%
TOTAL	100.0%	100.0%

V=0.027

Women and men express equal levels of satisfaction with their jobs (V = .03). This finding does not imply that women avoid the workplace problems just mentioned. Evidently when they answer the GSS question, they think of other aspects of their jobs.

If gender doesn't affect job satisfaction, what about race? To keep things simple, let's compare whites and African Americans. The latter have lower-paying jobs than whites and sometimes experience racial slights in the workplace.

<table>
<tr><td align="right">Data File:</td><td>GSS</td></tr>
<tr><td align="right">Task:</td><td>Cross-tabulation</td></tr>
<tr><td align="right">Row Variable:</td><td>10) LIKE JOB?</td></tr>
<tr><td align="right">➤ Column Variable:</td><td>42) WHTE/AFRAM</td></tr>
<tr><td align="right">➤ View:</td><td>Tables</td></tr>
<tr><td align="right">➤ Display:</td><td>Column %</td></tr>
</table>

42) WHTE/AFRAM

10) LIKE JOB?	WHITE	AFRICANA
VERY SAT.	46.4%	39.8%
MODER.SAT.	39.7%	38.9%
LIT./VERY	13.9%	21.3%
TOTAL	100.0%	100.0%

V=0.077**

African Americans are more likely (21.3 percent) to be a little dissatisfied or very dissatisfied with their jobs than white Americans (13.9 percent). The difference is small but statistically significant (V = .08**).

If educated people have higher-prestige jobs, as we saw in Exercise 11, should college-educated people express higher job satisfaction than people without a high school degree?

<table>
<tr><td align="right">Data File:</td><td>GSS</td></tr>
<tr><td align="right">Task:</td><td>Cross-tabulation</td></tr>
<tr><td align="right">Row Variable:</td><td>10) LIKE JOB?</td></tr>
<tr><td align="right">➤ Column Variable:</td><td>6) EDUCATION</td></tr>
<tr><td align="right">➤ View:</td><td>Tables</td></tr>
<tr><td align="right">➤ Display:</td><td>Column %</td></tr>
</table>

6) EDUCATION

10) LIKE JOB?	NO HS GRA	HS GRAD	SOME COL	COLL GRA
VERY SAT.	45.9%	43.4%	42.9%	48.2%
MODER.SAT.	38.4%	40.1%	42.6%	37.2%
LIT./VERY	15.7%	16.5%	14.5%	14.6%
TOTAL	100.0%	100.0%	100.0%	100.0%

V=0.037

This time our hypothesis is not supported (V = .04). Why not?

Let's change our focus to unemployment. A vast amount of literature documents the psychological and other effects of unemployment on individuals and their families. We can explore a few of these effects with GSS data by comparing those who are currently working with those who are currently unemployed. For now, let's look just at happiness.

<table>
<tr><td>Data File:</td><td>GSS</td></tr>
<tr><td>Task:</td><td>Cross-tabulation</td></tr>
<tr><td>➤ Row Variable:</td><td>132) HAPPY?</td></tr>
<tr><td>➤ Column Variable:</td><td>1) WORKING?</td></tr>
<tr><td>➤ View:</td><td>Tables</td></tr>
<tr><td>➤ Display:</td><td>Column %</td></tr>
</table>

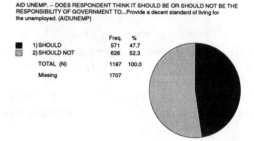

1) WORKING?		
	WORKING	UNEMPLOY
VERY HAPPY	29.3%	16.0%
PRET.HAPPY	60.2%	53.3%
NOT TOO	10.5%	30.7%
TOTAL	100.0%	100.0%

V=0.124**

Unemployed respondents are about three times as likely as working ones to say they're not too happy (V = .12**).

When it comes to helping the unemployed, the public is about evenly divided. The GSS asks whether it should be the government's responsibility "to provide a decent standard of living for the unemployed."

<table>
<tr><td>Data File:</td><td>GSS</td></tr>
<tr><td>➤ Task:</td><td>Univariate</td></tr>
<tr><td>➤ Primary Variable:</td><td>23) AID UNEMP.</td></tr>
<tr><td>➤ View:</td><td>Pie</td></tr>
</table>

AID UNEMP. -- DOES RESPONDENT THINK IT SHOULD BE OR SHOULD NOT BE THE RESPONSIBILITY OF GOVERNMENT TO...Provide a decent standard of living for the unemployed. (AIDUNEMP)

	Freq.	%
1) SHOULD	571	47.7
2) SHOULD NOT	626	52.3
TOTAL (N)	1197	100.0
Missing	1707	

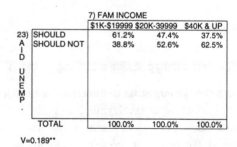

About 48 percent think the government should help the unemployed in this manner.

Should family income affect views on such aid?

<table>
<tr><td>Data File:</td><td>GSS</td></tr>
<tr><td>➤ Task:</td><td>Cross-tabulation</td></tr>
<tr><td>➤ Row Variable:</td><td>23) AID UNEMP.</td></tr>
<tr><td>➤ Column Variable:</td><td>7) FAM INCOME</td></tr>
<tr><td>➤ View:</td><td>Tables</td></tr>
<tr><td>➤ Display:</td><td>Column %</td></tr>
</table>

7) FAM INCOME		
	$1K-$19999	$20K-39999	$40K & UP
SHOULD	61.2%	47.4%	37.5%
SHOULD NOT	38.8%	52.6%	62.5%
TOTAL	100.0%	100.0%	100.0%

V=0.189**

Low-income people are more likely (61.2 percent) than high-income people (37.5 percent) to think the government should provide a decent standard of living for the unemployed (V = .19**). How would you explain this finding?

WORKSHEET

NAME:

COURSE:

DATE:

EXERCISE

12

REVIEW QUESTIONS

Based on the first part of this exercise, answer True or False to the following items:

In no country of the world is more than 60 percent of the GDP accounted for by agriculture.	T	F
Agricultural nations have higher infant mortality rates than industrial nations.	T	F
In the United States, the southeastern states have the highest proportion of people employed in agriculture.	T	F
In the United States, a state's degree of "agriculturalism" is not related to its level of education.	T	F
In the GSS, women are less satisfied than men with their jobs.	T	F
In the GSS, unemployed people feel less happy than working people.	T	F

EXPLORIT QUESTIONS

1. In the first part of this exercise you used the GSS to examine the relationship between happiness and one's employment status. At the international level, let's see if happiness is lower in nations with higher unemployment.

> ➤ Data File: **NATIONS**
> ➤ Task: **Scatterplot**
> ➤ Dependent Variable: **126) VERY HAPPY**
> ➤ Independent Variable: **30) UNEMPLYRT**
> ➤ View: **Reg. Line**

a. What is the value of r? $r =$ _____

b. Is r statistically significant? Yes No

c. Do these results surprise you? Explain why or why not in light of the results you found earlier for the GSS.

2. In the United States, do the manufacturing states have higher or lower levels of unemployment?

> ➤ *Data File:* **STATES**
> ➤ *Task:* **Scatterplot**
> ➤ *Dependent Variable:* **67) UNEMPLOYED**
> ➤ *Independent Variable:* **72) % MANUF.EMP**
> ➤ *View:* **Reg. Line**

Fill in the following sentence: The higher the percent of a state's workers employed in manufacturing, the _____ the state's unemployment rate.

3. Many people today find it difficult to juggle home and work responsibilities. The GSS asks, "How successful do you feel at balancing your paid work and your family life?"

> ➤ *Data File:* **GSS**
> ➤ *Task:* **Cross-tabulation**
> ➤ *Row Variable:* **83) BAL WK/FAM**
> ➤ *Column Variable:* **45) GENDER**
> ➤ *Subset Variable:* **54) MARITAL**
> ➤ *Subset Categories:* **Include: 1) MARRIED**
> ➤ *View:* **Tables**
> ➤ *Display:* **Column %**

> **The option for selecting a subset variable is located on the same screen you use to select other variables. For this example, select 54) MARITAL as a subset variable. A window will appear that shows you the categories of the subset variable. Select 1) Married as your subset category and choose the [Include] option. Then click [OK] and continue as usual.**

a. What percent of married women feel successful in balancing work and family life?   _____%

b. What percent of married men feel successful in balancing work and family life?   _____%

c. Is V statistically significant?   Yes   No

d. Does gender influence the extent to which we feel successful in balancing work and family life?   Yes   No

4. The GSS also asks whether respondents have ever missed a family occasion or holiday because of their work responsibilities.

> | | |
> |---:|:---|
> | Data File: | **GSS** |
> | Task: | **Cross-tabulation** |
> | ➤ Row Variable: | **98) MISS HOLS?** |
> | ➤ Column Variable: | **45) GENDER** |
> | ➤ Subset Variable: | **54) MARITAL** |
> | ➤ Subset Categories: | **Include: 1) MARRIED** |
> | ➤ View: | **Tables** |
> | ➤ Display: | **Column %** |

   The subset selection from the previous analysis continues until you exit the task, delete all subset variables, or clear all variables.

   a. What percent of women have missed a family event because of their
       job responsibilities?   _____%

   b. What percent of men have missed a family event because of their
       job responsibilities?   _____%

   c. Is V statistically significant?   Yes   No

   d. Does gender influence whether we miss family events because of our jobs?   Yes   No

5. Look back at your responses to these last two cross-tabulations. Are the findings in the two cross-tabulations consistent with each other? Explain your answer.

6. Affirmative action for women and minorities continues to be a controversial issue in the United States. The GSS asks respondents whether they are "for or against the preferential hiring and promotion of women."

> | | |
> |---:|:---|
> | Data File: | **GSS** |
> | ➤ Task: | **Univariate** |
> | ➤ Primary Variable: | **16) FEM JOB+** |
> | ➤ Display: | **Pie** |

  a. What percent of the sample is for affirmative action for women?   _____%

  b. What percent of the sample is against affirmative action for women?   _____%

  c. Are you for or against affirmative action for women? Explain why you feel the way you do.

7. Let's see whether gender affects views on affirmative action for women. What do you think we'll find?

> *Data File:* **GSS**
> ➤ *Task:* **Cross-tabulation**
> ➤ *Row Variable:* **16) FEM JOB+**
> ➤ *Column Variable:* **45) GENDER**
> ➤ *View:* **Tables**
> ➤ *Display:* **Column %**

  a. What percent of women are for affirmative action for women?   _____%

  b. What percent of men are for affirmative action for women?   _____%

  c. Is V statistically significant?   Yes   No

  d. Are women more likely than men to support affirmative action for women?   Yes   No

  e. Did this result surprise you? Why or why not?

8. The NATIONS data set includes a variable on the percent of children who are seriously underweight. We've seen that people in agricultural nations have some very negative life chances. Are their children especially likely to be underweight? To find out, please do the following:

a. In the NATIONS data set, map 25) %STUNTED. Note which part of the world generally has the highest proportions of underweight children.

b. Obtain the rankings of the nations on this variable, note which three nations have the highest proportions of underweight children, and the actual percent of their children who are underweight.

c. Next, obtain a scatterplot where 25) % STUNTED is the dependent variable and 33) % IN AGR. is the independent variable. This will allow you to determine whether the most agricultural nations tend to have the highest proportions of underweight children.

d. Using all the results you've obtained, write a brief essay below in which you describe the distribution of underweight children throughout the world and comment on the question of whether they are especially likely to be found in agricultural nations.

9. Have you ever wondered what life would be like if you won a state lottery? In particular, have you wondered whether you would continue to work if you didn't have to? The GSS asks respondents, "If you were to get enough money to live as comfortably as you would like for the rest of your life, would you continue to work or would you stop working?"

a. How would you answer this question?

I would continue to work

I would stop working

I'm not sure what I would do

b. Explain the answer you just gave.

c. About what percent of the GSS respondents do you think will say they would continue to work?

_____%

d. Obtain a univariate distribution for 11) WORK IF $$ to see how accurate your guess was. What percent of the GSS sample says it would continue to work?

_____%

e. Would you say your guess was very close, somewhat close, or not too close?

Very close

Somewhat close

Not too close

10. Do you think gender will affect whether people would want to continue to work? Why or why not?

a. Let's see whether your hypothesis was correct. First obtain the following cross-tabulation.

> Data File: **GSS**
> ➤ Task: **Cross-tabulation**
> ➤ Row Variable: **11) WORK IF $$**
> ➤ Column Variable: **45) GENDER**
> ➤ View: **Tables**
> ➤ Display: **Column %**

Discovering Sociology

b. Now answer these questions.

What percent of women say they would continue to work?   _____%

What percent of men?   _____%

Is V statistically significant?   Yes   No

Did the data support your hypothesis?   Yes   No

11. Finally, let's see whether family income predicts whether people would want to work if they didn't have to for economic reasons.

> Data File: **GSS**
> Task: **Cross-tabulation**
> Row Variable: **11) WORK IF $$**
> ➤ Column Variable: **7) FAM INCOME**
> ➤ View: **Tables**
> ➤ Display: **Column %**

In your own words, summarize the results of this analysis. Taking into account statistical significance, discuss whether this table supports the hypothesis that family income is related to whether people would want to continue to work if they didn't have to.

POLITICS AND GOVERNMENT

Tasks: Mapping, Cross-tabulation, Scatterplot, Historical Trends, Univariate
Data Files: NATIONS, STATES, GSS, HISTORY

Nations around the world differ in the types of government that run them. Some, like the United States, are democracies, in which people elect their representatives and political freedom is generally the rule. Others are more authoritarian; in these, elections are rare or are "sham" elections if they do occur, and one person or a small group of self-appointed individuals rules the government. Political freedom is uncommon and, even worse, political repression the norm; people live in fear for their lives if they dare question the arbitrary power under which they live.

Some of the most important questions in the study of politics and society concern the reasons why ordinary citizens become involved, or fail to become involved, in politics. In democracies, the most common political activity is voting, and many studies on voting exist. As we'll be seeing, some types of people are more likely to vote than others. In both democratic and authoritarian nations, citizens often try to influence the political process through nonelectoral means, including protest. Many types of protest exist, and we'll be looking at a few of them.

Our exploration begins with a look at political structures, activities, and attitudes across the world and within the United States. As you read through this exercise, think about why you have the attitudes you do about politics and government and why you've voted or engaged in some other political activity or, perhaps, failed to vote or otherwise become involved in politics. The exercise will give you a better idea of the social forces affecting your political behavior and attitudes.

GLOBAL POLITICS

The world's nations have been ranked according to how much political freedom their citizens enjoy. In the freest or most democratic nations, people vote via a secret ballot and enjoy freedom of speech, other political freedoms, and civil liberties. In the least free or most authoritarian nations, none of these hallmarks of democracy exist. In the NATIONS data set, the most democratic nations get a score of 7 and the most authoritarian get a score of 1. Let's see which nations are the most democratic and which are the most authoritarian.

> *Data File:* **NATIONS**
> *Task:* **Mapping**
> *Variable 1:* **57) DEMOCRACY**
> *View:* **Map**

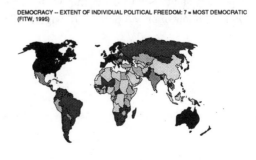

DEMOCRACY -- EXTENT OF INDIVIDUAL POLITICAL FREEDOM: 7 = MOST DEMOCRATIC
(FITW, 1995)

The darker the color, the more democratic. As you might have expected, the most democratic nations tend to be in North America and Western Europe.

An important variable to political sociologists is political interest, which, as the name implies, concerns the amount of interest someone has in politics. The NATIONS data set includes a measure of the percent of each nation's respondents who are very interested or somewhat interested in politics.

Data File: **NATIONS**
Task: **Mapping**
➤ Variable 1: **86) P. INTEREST**
➤ View: **Map**

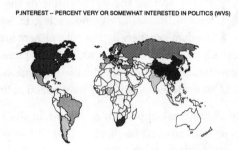

P.INTEREST -- PERCENT VERY OR SOMEWHAT INTERESTED IN POLITICS (WVS)

The amount of political interest varies around the world, but generally appears highest in Europe.

We said earlier that people sometimes take part in nonelectoral activity to influence the political process. The NATIONS data set includes three such activities: signing a petition, taking part in a boycott, and participating in a lawful demonstration.

Data File: **NATIONS**
Task: **Mapping**
➤ Variable 1: **87) PETITION?**
➤ View: **List: Rank**

RANK	CASE NAME	VALUE
1	Canada	77
2	United Kingdom	75
3	Sweden	72
4	United States	71
5	Latvia	65
6	Switzerland	63
7	Japan	62
8	Norway	61
9	Germany	60
10	Lithuania	58

The percent of citizens who say they've signed a political petition ranges from a high of 77 percent in Canada to a low of only 5 percent in Nigeria. The United States ranks somewhat below Canada, at 71 percent.

Discovering Sociology

Data File:	**NATIONS**		
Task:	**Mapping**		
➤ Variable 1:	**88) BOYCOTT?**		
➤ View:	**List: Rank**		

RANK	CASE NAME	VALUE
1	South Africa	22
1	Canada	22
3	Iceland	21
4	United States	18
5	Sweden	17
6	United Kingdom	15
7	Finland	14
8	Italy	13
8	France	13
10	Norway	12

The percent of citizens who say they've joined a boycott is much lower than the percent of those who have signed a petition. Boycotting ranges from a high of 22 percent in South Africa and Canada to a low of only 2 percent in Hungary. The United States again ranks below Canada, at 18 percent.

POLITICS IN THE UNITED STATES

In the 1992 presidential election, Democratic Party candidate Bill Clinton defeated Republican Party incumbent George Bush with Ross Perot running a strong third. Let's see where in the country Clinton's support was highest and then see, demographically speaking, what types of states gave him the most and the least support.

➤ Data File: **STATES**
➤ Task: **Mapping**
➤ Variable 1: **106) %CLINTON92**
➤ View: **Map**

%CLINTON92 -- 1992: PERCENT OF VOTES FOR CLINTON (DEM.) (WA, 1993)

The percent voting for Clinton was generally highest in California, in New Mexico, and east of the Mississippi.

	Data File:	**STATES**
	Task:	**Mapping**
	Variable 1:	**106) %CLINTON92**
➤	View:	**List: Rank**

RANK	CASE NAME	VALUE
1	Arkansas	54
2	Maryland	50
2	New York	50
4	West Virginia	49
4	Hawaii	49
6	Rhode Island	48
6	Massachusetts	48
6	Illinois	48
9	Tennessee	47
9	California	47

It ranged from a high of 54 percent in Arkansas, his home state, to a low of 26 percent in Utah.

Why should some states be more likely than others to favor Clinton? Historically, the Democratic Party has been seen as the party of the working class, whereas the Republican Party has been seen as the party of wealthier folk. If that's the case, states with more lower-income families should have been more likely to vote for Clinton.

	Data File:	**STATES**
➤	Task:	**Scatterplot**
➤	Dependent Variable:	**106) %CLINTON92**
➤	Independent Variable:	**65) MED.FAM. $**
➤	View:	**Reg. Line**

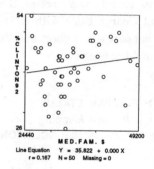

There's a slight positive correlation here (r = .17), opposite to what we expected, that's not statistically significant. The states with more lower-income families were not more likely to vote for Clinton.

What about unemployment? The U.S. economy in the early 1990s was stagnant, and many people were worried about their economic future. Perhaps the states with higher unemployment rates were more likely to vote for Clinton.

Discovering Sociology

Data File: **STATES**
Task: **Scatterplot**
Dependent Variable: **106) %CLINTON92**
➤ Independent Variable: **67) UNEMPLOYED**
➤ View: **Reg. Line**

Line Equation   Y = 31.942 + 1.495 X
r = 0.385**   N = 50   Missing = 0

Our hypothesis is supported (r = .39**).

Since the presidency of Franklin Delano Roosevelt some 60 years ago, African Americans have supported the Democratic Party in great numbers. Perhaps the states with the highest proportions of African Americans were more likely to vote for Clinton.

Data File: **STATES**
Task: **Scatterplot**
Dependent Variable: **106) %CLINTON92**
➤ Independent Variable: **14) %AFRIC.AM**
➤ View: **Reg. Line**

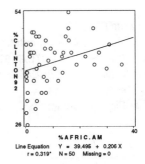

Line Equation   Y = 39.495 + 0.206 X
r = 0.319*   N = 50   Missing = 0

Our hypothesis is again supported (r = .32*).

We now turn to the GSS and continue our look at voting. Let's first get an idea of what kinds of people were more and less likely to vote in 1992. We'll start with education. One of the most consistent findings in the voting literature is that more-educated people vote more regularly than less-educated people. Let's find out whether the GSS data reflect this difference.

➤ Data File: **GSS**
➤ Task: **Cross-tabulation**
➤ Row Variable: **100) VOTE IN 92**
➤ Column Variable: **6) EDUCATION**
➤ View: **Tables**
➤ Display: **Column %**

6) EDUCATION

100) VOTE IN 92	NO HS GRA	HS GRAD	SOME COL	COLL GRA
VOTED	51.7%	65.6%	74.1%	86.7%
DID NOT	48.3%	34.4%	25.9%	13.3%
TOTAL	100.0%	100.0%	100.0%	100.0%

V=0.261**

People with a college degree are more likely (86.7 percent) than those without a high school degree (51.7 percent) to report voting in 1992 (V = .26**).

Data File: **GSS**
Task: **Cross-tabulation**
Row Variable: **100) VOTE IN 92**
➤ Column Variable: **7) FAM INCOME**
➤ View: **Tables**
➤ Display: **Column %**

7) FAM INCOME			
100) VOTE IN 92	$1K-$19999	$20K-39999	$40K & UP
VOTED	60.6%	71.8%	78.9%
DID NOT	39.4%	28.2%	21.1%
TOTAL	100.0%	100.0%	100.0%

V=0.164**

Perhaps reflecting the education difference, the highest-income group is more likely than the lowest to report voting (V = .16**).

What difference, if any, did race and gender make?

Data File: **GSS**
Task: **Cross-tabulation**
Row Variable: **100) VOTE IN 92**
➤ Column Variable: **43) RACE/ETHNC**
➤ View: **Tables**
➤ Display: **Column %**

43) RACE/ETHNC					
100) VOTE IN 92	WHITE	AFRICANA	LATINO	NATIVEAM	ASIAN/PAC
VOTED	74.1%	67.5%	51.9%	54.5%	41.2%
DID NOT	25.9%	32.5%	48.1%	45.5%	58.8%
TOTAL	100.0%	100.0%	100.0%	100.0%	100.0%

V=0.130**

Race is related to voting (V = .13**): whites were slightly more likely than African Americans to report voting in 1992, and African Americans were more likely than Latinos or Native Americans. The Asian-American figure, 41.2 percent, is lower yet, but is based on only 17 people, too low a number to draw a reliable conclusion from.

Data File: **GSS**
Task: **Cross-tabulation**
Row Variable: **100) VOTE IN 92**
➤ Column Variable: **45) GENDER**
➤ View: **Tables**
➤ Display: **Column %**

45) GENDER		
100) VOTE IN 92	FEMALE	MALE
VOTED	72.3%	70.3%
DID NOT	27.7%	29.7%
TOTAL	100.0%	100.0%

V=0.023

No gender difference in voting appears (V = .02).

Were the politically alienated least likely to vote? We'll look at agreement with the statement "People like me don't have any say about what the government does."

Data File: **GSS**
Task: **Cross-tabulation**
Row Variable: **100) VOTE IN 92**
➤ Column Variable: **144) POL.EFF.11**
➤ View: **Tables**
➤ Display: **Column %**

144) POL.EFF.11

100) VOTE IN 92	AGREE	NEITHER	DISAGREE
VOTED	65.4%	67.2%	80.7%
DID NOT	34.6%	32.8%	19.3%
TOTAL	100.0%	100.0%	100.0%

V=0.161**

The alienated ("agree") are less likely to report voting in 1992 (V = .16**).

If they did vote, were they especially likely to vote for Perot?

Data File: **GSS**
Task: **Cross-tabulation**
➤ Row Variable: **101) WHO IN 92?**
➤ Column Variable: **144) POL.EFF.11**
➤ View: **Tables**
➤ Display: **Column %**

144) POL.EFF.11

101) WHO IN 92?	AGREE	NEITHER	DISAGREE
CLINTON	53.0%	51.3%	44.5%
BUSH	31.2%	39.3%	42.3%
PEROT	15.9%	9.4%	13.3%
TOTAL	100.0%	100.0%	100.0%

V=0.084*

The alienated ("agree") were only slightly more likely than the nonalienated ("disagree") to vote for Perot. Since Clinton's share of the "alienated" vote was higher than his share of the other two categories, would you say he benefited the most from political alienation in 1992 (V = .08*)?

What other factors affected candidate preferences in 1992? Let's try family income.

Data File: **GSS**
Task: **Cross-tabulation**
Row Variable: **101) WHO IN 92?**
➤ Column Variable: **7) FAM INCOME**
➤ View: **Tables**
➤ Display: **Column %**

7) FAM INCOME

101) WHO IN 92?	$1K-$19999	$20K-39999	$40K & UP
CLINTON	60.6%	49.1%	42.5%
BUSH	28.0%	33.0%	41.2%
PEROT	11.4%	18.0%	16.3%
TOTAL	100.0%	100.0%	100.0%

V=0.106**

Income is very much related to candidate preference (V = .11**). The lowest-income group was more likely than the other income groups to vote for Clinton. Conversely, the highest-income group was more likely to vote for Bush. The lowest-income group gave Perot the least support.

What about race? Here we'll just compare whites and African Americans.

Data File: **GSS**
Task: **Cross-tabulation**
Row Variable: **101) WHO IN 92?**
➤ Column Variable: **42) WHTE/AFRAM**
➤ View: **Tables**
➤ Display: **Column %**

42) WHTE/AFRAM

101) WHO IN 92?	WHITE	AFRICANA
CLINTON	41.6%	92.4%
BUSH	41.1%	4.6%
PEROT	17.3%	2.9%
TOTAL	100.0%	100.0%

V=0.347**

African Americans overwhelmingly voted for Clinton (V = .35**).

Have you heard of the gender gap in politics? Let's see how much of a gender gap there was in the 1992 election.

Data File: **GSS**
Task: **Cross-tabulation**
Row Variable: **101) WHO IN 92?**
➤ Column Variable: **45) GENDER**
➤ View: **Tables**
➤ Display: **Column %**

45) GENDER

101) WHO IN 92?	FEMALE	MALE
CLINTON	55.0%	41.9%
BUSH	33.0%	39.3%
PEROT	12.0%	18.9%
TOTAL	100.0%	100.0%

V=0.138**

Women were more likely (55.0 percent) than men (41.9 percent) to vote for Clinton (V = .14**).

Let's turn to political ideology and see whether Americans have become more conservative in the last quarter century.

➤ Data File: **HISTORY**
➤ Task: **Historical Trends**
➤ Variable: **17) POL. VIEW**

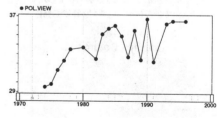

Percent saying their views are conservative

Americans are slightly more conservative than they were 25 years ago.

Several factors may influence how liberal or conservative people are. Let's look first at gender.

> ➤ Data File: **GSS**
> ➤ Task: **Cross-tabulation**
> ➤ Row Variable: **102) POL. VIEW**
> ➤ Column Variable: **45) GENDER**
> ➤ View: **Tables**
> ➤ Display: **Column %**

45) GENDER		
102) POL. VIEW	FEMALE	MALE
LIBERAL	26.9%	23.5%
MODERATE	39.3%	36.6%
CONSERV.	33.8%	39.9%
TOTAL	100.0%	100.0%

V=0.064**

Women are slightly less conservative than men (V = .06**).

Do you think African Americans will be more liberal than whites?

> Data File: **GSS**
> Task: **Cross-tabulation**
> Row Variable: **102) POL. VIEW**
> ➤ Column Variable: **42) WHTE/AFRAM**
> ➤ View: **Tables**
> ➤ Display: **Column %**

42) WHTE/AFRAM		
102) POL. VIEW	WHITE	AFRICANA
LIBERAL	24.3%	30.9%
MODERATE	38.1%	36.2%
CONSERV.	37.6%	32.9%
TOTAL	100.0%	100.0%

V=0.054*

African Americans are slightly more likely than whites to say they're liberal (V = .05*).

Now let's look at family income.

> Data File: **GSS**
> Task: **Cross-tabulation**
> Row Variable: **102) POL. VIEW**
> ➤ Column Variable: **7) FAM INCOME**
> ➤ View: **Tables**
> ➤ Display: **Column %**

7) FAM INCOME			
102) POL. VIEW	$1K-$19999	$20K-39999	$40K & UP
LIBERAL	26.8%	26.0%	25.4%
MODERATE	40.6%	39.2%	34.7%
CONSERV.	32.6%	34.7%	39.9%
TOTAL	100.0%	100.0%	100.0%

V=0.048*

The wealthier the respondent, the more conservative (V = .05*).

Finally, let's see whether religiosity is related to political ideology. What do you think we'll find?

Data File: **GSS**
Task: **Cross-tabulation**
Row Variable: **102) POL. VIEW**
➤ Column Variable: **127) ATTEND**
➤ View: **Tables**
➤ Display: **Column %**

	127) ATTEND		
	NEVER	MONTH/YR	WEEKLY
102) LIBERAL	32.4%	28.4%	16.8%
P MODERATE	38.2%	40.0%	35.5%
O CONSERV.	29.4%	31.6%	47.7%
L			
.			
V			
I			
E			
W			
TOTAL	100.0%	100.0%	100.0%

V=0.124**

People who never attend religious services are about twice as likely as those who attend weekly to be liberal (V = .12**).

WORKSHEET

NAME:

COURSE:

DATE:

EXERCISE
13

REVIEW QUESTIONS

Based on the first part of this exercise, answer True or False to the following items:

The most democratic nations tend to be in North America and Western Europe.	T	F
Political interest is highest in North America.	T	F
In the United States, the midwestern states were the most likely to vote for Bill Clinton in 1992.	T	F
The poorest states were especially likely to vote for Clinton in 1992.	T	F
In the GSS, African Americans were much more likely than whites to vote for Clinton in 1992.	T	F
In the GSS, men were more likely than women to vote for Clinton in 1992.	T	F

EXPLORIT QUESTIONS

1. The NATIONS data set includes a measure of the percent of people identifying themselves as being on the political left.

    ➤ *Data File:* **NATIONS**
    ➤ *Task:* **Mapping**
    ➤ *Variable 1:* **83) % LEFTISTS**
    ➤ *View:* **List: Rank**

    a. Which nation has the highest percent of people identifying themselves as being on the political left? _____

    b. Which nation has the lowest percent of people identifying themselves as being on the political left? _____

    c. What is the percent for the United States? _____%

2.   Did Perot's vote in 1992 reflect a state's proportion of African Americans?

> ➤ *Data File:* **STATES**
> ➤ *Task:* **Scatterplot**
> ➤ *Dependent Variable:* **108) % PEROT 92**
> ➤ *Independent Variable:* **14) %AFRIC.AM**
> ➤ *View:* **Reg. Line**

a.   What is the value of r?   r = _____

b.   Is r statistically significant?   Yes   No

c.   In 1992, did a state's African-American population predict its percent voting for Perot?   Yes   No

3.   The GSS asks whether the federal government "has too much power or too little power." Given the presence of antigovernment militias in the western part of the nation, our hypothesis will be that the West will be more likely than other regions to say the government has too much power.

> ➤ *Data File:* **GSS**
> ➤ *Task:* **Cross-tabulation**
> ➤ *Row Variable:* **112) GOV. POW.**
> ➤ *Column Variable:* **62) REGION**
> ➤ *View:* **Tables**
> ➤ *Display:* **Column %**

a.   What percent of the West thinks the government has too much power?   _____%

b.   Is V statistically significant?   Yes   No

c.   What would you tell someone if she/he said that the presence of so many antigovernment militias in the West reflects the general antigovernment sentiment that is found there?

4. One of the hallmarks of democracy is whether people are freely allowed to express their views, no matter how unpopular they may be. The GSS asks whether an atheist should be allowed to make a speech in the respondent's community against churches and religion.

> Data File: **GSS**
> Task: **Cross-tabulation**
> ➤ Row Variable: **103) ATHEIST SP**
> ➤ Column Variable: **127) ATTEND**
> ➤ View: **Tables**
> ➤ Display: **Column %**

a. What percent of people who never attend religious services think an atheist should be allowed to speak? _____%

b. What percent of people who attend religious services weekly think an atheist should be allowed to speak? _____%

c. Is V statistically significant?   Yes   No

d. Does religiosity influence the extent to which we believe in freedom of speech for atheists? Explain your answer.

5. The GSS asked a similar question as to whether a racist claiming African Americans are genetically inferior should be allowed to speak.

> Data File: **GSS**
> Task: **Cross-tabulation**
> ➤ Row Variable: **104) RACIST SP**
> ➤ Column Variable: **42) WHTE/AFRAM**
> ➤ View: **Tables**
> ➤ Display: **Column %**

a. What percent of whites think a racist should be allowed to speak? _____%

b. What percent of African Americans think a racist should be allowed to speak? _____%

   c. Is V statistically significant?   Yes   No

   d. Does race influence the extent to which we believe in freedom of speech for racists? Explain your answer.

6.   Now for a similar question about allowing a Communist to speak.

> | | |
> |---:|:---|
> | *Data File:* | **GSS** |
> | *Task:* | **Cross-tabulation** |
> | ➤ *Row Variable:* | **105) COMMUN.SP** |
> | ➤ *Column Variable:* | **102) POL. VIEW** |
> | ➤ *View:* | **Tables** |
> | ➤ *Display:* | **Column %** |

   a. What percent of liberals think a Communist should be allowed to speak?   _____%

   b. What percent of conservatives think a Communist should be allowed to speak?   _____%

   c. Is V statistically significant?   Yes   No

   d. Does our political ideology influence the extent to which we believe in freedom of speech for Communists? Explain your answer.

7.   Finally, let's see whether views about homosexuality are related to whether people think a homosexual should be allowed to speak.

> | | |
> |---:|:---|
> | *Data File:* | **GSS** |
> | *Task:* | **Cross-tabulation** |
> | ➤ *Row Variable:* | **106) HOMO. SP** |
> | ➤ *Column Variable:* | **74) HOMO. SEX** |
> | ➤ *View:* | **Tables** |
> | ➤ *Display:* | **Column %** |

a. What percent of people who think homosexuality is always wrong think
   a homosexual should be allowed to speak?   _____%

b. What percent of people who think homosexuality is not wrong think
   a homosexual should be allowed to speak??   _____%

c. Is V statistically significant?   Yes   No

d. Do views about homosexuality influence the extent to which we believe in freedom of speech for
   homosexuals? Explain your answer

8. Looking at your results for all the analyses of freedom of speech, what general conclusion(s) can you
   draw?

9. We hear a lot of criticism these days about the federal government. Some of it is directed at the White
   House, and other criticism is directed at the Congress. The U.S. Supreme Court also comes in for its
   share of criticism. The GSS asks respondents how much confidence they have in each of these three
   branches of government. Complete each of the following analyses and keep notes on the results.

> Data File: **GSS**
> ➤ Task: **Univariate**
> ➤ Primary Variable: **107) FED.GOV'T**
> ➤ View: **Pie**

> Data File: **GSS**
> Task: **Univariate**
> ➤ Primary Variable: **109) CONGRESS?**
> ➤ View: **Pie**

> Data File: **GSS**
> Task: **Univariate**
> ➤ Primary Variable: **108) SUP.COURT?**
> ➤ View: **Pie**

a. Generally speaking, in which branch of the government do people have the most confidence? (Circle one)?

Executive branch

Congress

Supreme Court

b. What do you think accounts for the difference you just noted? Why do you think confidence in the other two branches is so low?

10. Finally, let's see how confidence in the White House and Congress has changed in the past 25 years. We'll examine changes in the percent of the GSS reporting a "great deal of confidence" in these two branches of the government.

> ➤ *Data File:* **HISTORY**
>   ➤ *Task:* **Historical Trends**
> ➤ *Variable 1:* **18) FED.GOVT**
> ➤ *Variable 2:* **19) CONGRESS**

a. Generally speaking, would you say that the trends for the White House and Congress are similar to each other, or are they very different from each other? (Circle one)?

Similar

Very different

b. Compare the changes depicted in the graphs with the events listed below them. (Hint: click on the small lines that represent the years.) Does it seem as though the changes are related to any or all of the events? Choosing one or two events, write a brief essay that supports your view.

HEALTH, ILLNESS, AND MEDICINE

Tasks: Mapping, Scatterplot, Univariate, Cross-tabulation, Historical Trends
Data Files: NATIONS, STATES, GSS, HISTORY

Medicine is yet another institution that is very much a part of our lives. Most of us were born in hospitals, and many of us will die in hospitals or other health-care facilities. In between, we spend a lot of time in the doctor's office when we or our loved ones are sick or injured. We go to the pharmacy to get prescription medications and to any number of places to get over-the-counter products such as aspirin and cough medicine. One way or the other, medicine affects all of us.

As this discussion should suggest, health and illness are far more than the medical matters that most health professionals consider them. Sociologists and many public health professionals recognize that health and illness are social and cultural matters as well. As such, they reflect many of the other factors we've been examining throughout this workbook: social class, gender, race/ethnicity, age, religiosity, and so on. In some cases these effects stem from the physiological differences between, say, women and men or the elderly and the non-elderly. But in other cases they reflect the fault lines of a society where people with different socioeconomic rankings have different life outcomes.

This exercise explores some important aspects of health and medicine in today's world. We will look both at indicators of health and illness and at views on important health and medical issues. This focus will help reinforce the sociologist's emphasis on the social and cultural aspects of health and medicine.

A CROSS-CULTURAL LOOK AT HEALTH AND ILLNESS

When we examined stratification and the economy in previous exercises, we saw that the underdeveloped nations are far worse off than the developed nations on many indicators of health and illness. To reinforce this point, we map one such indicator, infant mortality, below. Notice the huge gap between North America and Western Europe and much of the rest of the world.

➤ *Data File:* **NATIONS**
➤ *Task:* **Mapping**
➤ *Variable 1:* **11) INF. MORTL**
➤ *View:* **Map**

INF. MORTL -- NUMBER OF INFANT DEATHS PER 1,000 BIRTHS (TWF 1994)

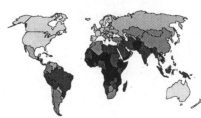

Ironically, however, some behaviors that put people at risk for illness and early death are more common in North America and Western Europe. Let's look at two of these risk factors.

Data File: **NATIONS**
Task: **Mapping**
➤ *Variable 1:* **124) CIGARETTES**
➤ *View:* **Map**

CIGARETTES -- CIGARETTE CONSUMPTION IN NUMBER PER CAPITA (NBWR, 1991)

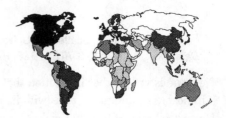

Cigarette smoking appears highest in North America, Europe, and parts of Asia. Despite their greater health problems in other respects, most of the underdeveloped nations have low rates of cigarette use.

Data File: **NATIONS**
Task: **Mapping**
Variable 1: **124) CIGARETTES**
➤ *View:* **List: Rank**

RANK	CASE NAME	VALUE
1	Albania	3446
2	Greece	2890
3	Japan	2603
4	United States	2567
5	Canada	2566
6	Iceland	2500
7	Hungary	2486
8	Poland	2484
9	Austria	2143
10	Switzerland	2117

Albania has the dubious honor of leading the world in cigarette smoking, with the United States and Canada virtually tied for fourth place.

Now let's look at alcohol use, which, if excessive, can lead to cirrhosis of the liver and other health problems.

Data File: **NATIONS**
Task: **Mapping**
➤ *Variable 1:* **120) ALCOHOL**
➤ *View:* **Map**

ALCOHOL -- NET ANNUAL ALCOHOL CONSUMPTION PER CAPITA, IN LITERS (IP)

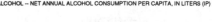
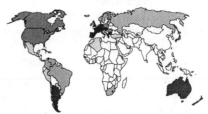

Alcohol use appears most common in Europe and least common in the less developed nations elsewhere.

HEALTH AND MEDICINE IN THE UNITED STATES

At the state level we can see regional patterns of medical use and of illness. Let's start with AIDS, using the number of AIDS deaths through 1991 per 100,000 population.

> ➤ *Data File:* **STATES**
> ➤ *Task:* **Mapping**
> ➤ *Variable 1:* **91) AIDS**
> ➤ *View:* **Map**

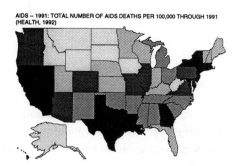

AIDS – 1991: TOTAL NUMBER OF AIDS DEATHS PER 100,000 THROUGH 1991 (HEALTH, 1992)

AIDS deaths are highest in the Northeast, Georgia, Louisiana, Texas, and California, and lowest in the upper Midwest and the mountain states.

Although many people associate AIDS with homosexuality, AIDS also results from the sharing of needles in heroin use and from unprotected heterosexual sex. Do you think all three behaviors are more common in urban areas than in rural areas? If so, the more urban states should have higher AIDS death rates.

> *Data File:* **STATES**
> ➤ *Task:* **Scatterplot**
> ➤ *Dependent Variable:* **91) AIDS**
> ➤ *Independent Variable:* **11) %URBAN**
> ➤ *View:* **Reg. Line**

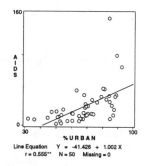

Line Equation   Y = -41.426 + 1.002 X
r = 0.555**   N = 50   Missing = 0

The more urban a state, the higher its AIDS death rate (r = .56**).

Being overweight is a health risk factor for many people. Which part of the country should be the heaviest?

Data File: **STATES**
➤ Task: **Mapping**
➤ Variable 1: **89) %FAT**
➤ View: **Map**

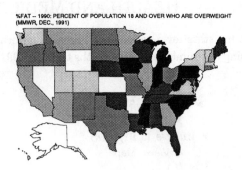

%FAT -- 1990: PERCENT OF POPULATION 18 AND OVER WHO ARE OVERWEIGHT (MMWR, DEC., 1991)

Obesity is more common east of the Mississippi. Why do you think this is so? Is it possible that more educated people are more aware of the health risk that obesity poses? If so, states with more college-educated people should have lower rates of obesity.

Data File: **STATES**
➤ Task: **Scatterplot**
➤ Dependent Variable: **89) %FAT**
➤ Independent Variable: **75) COLL.DEGR.**
➤ View: **Reg. Line**

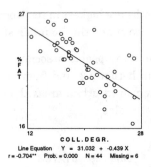

Line Equation   Y = 31.032 + -0.439 X
r = -0.704**   Prob. = 0.000   N = 44   Missing = 6

The scatterplot strongly supports the hypothesis (r = –.70**).

Should the more "overweight" states have higher rates of death by cardiovascular disease?

Data File: **STATES**
Task: **Scatterplot**
➤ Dependent Variable: **45) HEART DTHS**
➤ Independent Variable: **89) %FAT**
➤ View: **Reg. Line**

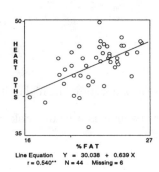

Line Equation   Y = 30.038 + 0.639 X
r = 0.540**   N = 44   Missing = 6

States with higher proportions of overweight people do have higher rates of cardiovascular deaths (r = .54**).

Many Americans lack health-care insurance. Let's see which parts of the country have the highest proportions of people without such insurance.

> *Data File:* **STATES**
> ➤ *Task:* **Mapping**
> ➤ *Variable 1:* **92) HEALTH.INS**
> ➤ *View:* **Map**

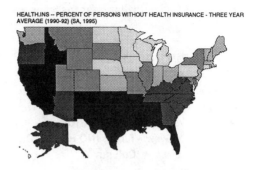

HEALTH.INS -- PERCENT OF PERSONS WITHOUT HEALTH INSURANCE - THREE YEAR AVERAGE (1990-92) (SA, 1995)

Should the poorer states have greater proportions of people without medical insurance?

> *Data File:* **STATES**
> ➤ *Task:* **Scatterplot**
> ➤ *Dependent Variable:* **92) HEALTH.INS**
> ➤ *Independent Variable:* **55) %POOR**
> ➤ *View:* **Reg. Line**

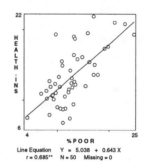

Line Equation   Y = 5.038 + 0.643 X
r = 0.685**   N = 50   Missing = 0

Most definitely (r = .69**).

We now continue our look at health and medicine with GSS data. We'll start with respondents' assessment of their own health.

> ➤ *Data File:* **GSS**
> ➤ *Task:* **Univariate**
> ➤ *Primary Variable:* **133) HEALTH**
> ➤ *View:* **Pie**

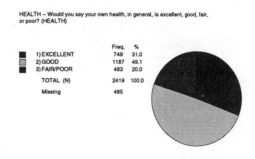

HEALTH -- Would you say your own health, in general, is excellent, good, fair, or poor? (HEALTH)

	Freq.	%
1) EXCELLENT	749	31.0
2) GOOD	1187	49.1
3) FAIR/POOR	483	20.0
TOTAL (N)	2419	100.0
Missing	485	

Almost one-third of the sample reports excellent health, and just under 50 percent reports good health. Twenty percent reports only fair or poor health.

In Exercise 6 we saw that poor people have worse health than the non-poor. If that's true only because the poor are less educated, then if we controlled for education, income should no longer be related to health. That is, people with the same education should not differ in health even if they differ in income. Let's find out what happens when we do control for education.

[Control: NO HS GRAD]

Data File: **GSS**
➤ Task: **Cross-tabulation**
➤ Row Variable: **133) HEALTH**
➤ Column Variable: **7) FAM INCOME**
➤ Control Variable: **6) EDUCATION**
➤ View: **Tables (NO HS GRAD)**
➤ Display: **Column %**

7) FAM INCOME			
133)		$1K-$19999	$20K-39999	$40K & UP
H E A L T H	EXCELLENT	12.7%	20.2%	20.0%
	GOOD	35.8%	47.5%	49.1%
	FAIR/POOR	51.5%	32.3%	30.9%
	TOTAL	100.0%	100.0%	100.0%

V=0.141**

The option for selecting a control variable is located on the same screen you use to select other variables. For this example, select **6) EDUCATION** as a control variable and then click **[OK]** to continue as usual. Separate tables for each of the 6) EDUCATION categories will now be shown for the 133) HEALTH and 7) FAM INCOME cross-tabulation. Examine these results before continuing.

[Control: HS GRAD]

➤ View: **Tables (HS GRAD)**
➤ Display: **Column %**

7) FAM INCOME			
133)		$1K-$19999	$20K-39999	$40K & UP
H E A L T H	EXCELLENT	21.1%	29.3%	30.9%
	GOOD	49.2%	50.2%	51.8%
	FAIR/POOR	29.6%	20.5%	17.3%
	TOTAL	100.0%	100.0%	100.0%

V=0.096*

Click the appropriate button at the bottom of the task bar to look at the second (or "next") partial table for 6) EDUCATION. Examine these results before continuing.

[Control: SOME COLL.]

➤ View: **Tables (SOME COLL.)**
➤ Display: **Column %**

7) FAM INCOME			
133)		$1K-$19999	$20K-39999	$40K & UP
H E A L T H	EXCELLENT	23.8%	31.7%	32.8%
	GOOD	48.3%	56.3%	60.9%
	FAIR/POOR	27.9%	12.1%	6.4%
	TOTAL	100.0%	100.0%	100.0%

V=0.177**

Again, click the appropriate button at the bottom of the task bar to look at the third (or "next") partial table for 6) EDUCATION. Examine these results before continuing.

➤ *View:* **Tables (COLL GRAD)**
➤ *Display:* **Column %**

	7) FAM INCOME		
	$1K-$19999	$20K-39999	$40K & UP
133) EXCELLENT	32.0%	41.6%	49.1%
H GOOD	49.3%	49.0%	43.9%
E FAIR/POOR	18.7%	9.4%	7.0%
A			
L			
T			
H			
TOTAL	100.0%	100.0%	100.0%

V=0.111**

Click the appropriate button at the bottom of the task bar to look at the last (or "next") partial table for 6) EDUCATION. This table includes only college graduates.

Even when we control for education, income continues to be related to health: the lower the income, the greater the likelihood of only fair or poor health. While lower education may be one of the reasons for the poor health of poor people, other factors, such as lack of medical insurance or access to good medical care, must also be at work.

The latter possibilities are illustrated when we look at another GSS variable asking whether respondents received medical treatment in a doctor's office, medical clinic, or hospital during the past week. Let's first see what percent were treated.

Data File: **GSS**
➤ *Task:* **Univariate**
➤ *Primary Variable:* **157) GO DOC**
➤ *View:* **Pie**

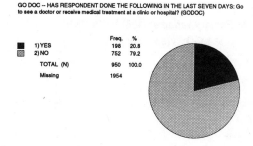

GO DOC -- HAS RESPONDENT DONE THE FOLLOWING IN THE LAST SEVEN DAYS: Go to see a doctor or receive medical treatment at a clinic or hospital? (GODOC)

		Freq.	%
■	1) YES	198	20.8
▨	2) NO	752	79.2
	TOTAL (N)	950	100.0
	Missing	1954	

Almost 21 percent of respondents have received medical care within the past week.

Now let's see whether people in worse health are more likely to seek medical treatment.

Data File: **GSS**
➤ *Task:* **Cross-tabulation**
➤ *Row Variable:* **157) GO DOC**
➤ *Column Variable:* **133) HEALTH**
➤ *View:* **Tables**
➤ *Display:* **Column %**

	133) HEALTH		
	EXCELLEN	GOOD	FAIR/POOR
157) YES	16.8%	17.9%	34.8%
G NO	83.2%	82.1%	65.2%
O			
D			
O			
C			
TOTAL	100.0%	100.0%	100.0%

V=0.170**

Not surprisingly, people in only fair or poor health are about twice as likely as people in better health to seek medical care (V = .17**).

Since lower-income people have worse health, they should be more likely, all other things equal, to get medical care. Let's see whether this is the case.

<table>
<tr><td>Data File:</td><td>GSS</td></tr>
<tr><td>Task:</td><td>Cross-tabulation</td></tr>
<tr><td>Row Variable:</td><td>157) GO DOC</td></tr>
<tr><td>➤ Column Variable:</td><td>7) FAM INCOME</td></tr>
<tr><td>➤ View:</td><td>Tables</td></tr>
<tr><td>➤ Display:</td><td>Column %</td></tr>
</table>

7) FAM INCOME			
	$1K-$19999	$20K-39999	$40K & UP
157) GO DOC YES	21.9%	18.0%	21.8%
NO	78.1%	82.0%	78.2%
TOTAL	100.0%	100.0%	100.0%

V=0.044

Despite the worse health of the poor, lower-income people are *not* more likely than higher-income people to get medical care (V = .04). While this finding supports the possibility that those with lower incomes can't afford medical care, other reasons for their lack of care, such as a sense of fatalism or lack of time, are also possible. How would you explain their lack of medical care?

As the United States continues to debate health-care policy, it's important to know public opinion on medicine and health care. The GSS asks how much confidence respondents have in medicine.

<table>
<tr><td>Data File:</td><td>GSS</td></tr>
<tr><td>➤ Task:</td><td>Univariate</td></tr>
<tr><td>➤ Primary Variable:</td><td>138) MEDICINE?</td></tr>
<tr><td>➤ View:</td><td>Pie</td></tr>
</table>

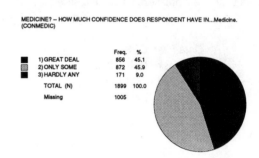

MEDICINE? -- HOW MUCH CONFIDENCE DOES RESPONDENT HAVE IN...Medicine. (CONMEDIC)

	Freq.	%
1) GREAT DEAL	856	45.1
2) ONLY SOME	872	45.9
3) HARDLY ANY	171	9.0
TOTAL (N)	1899	100.0
Missing	1005	

Forty-five percent of the public has a great deal of confidence in medicine, and 46 percent has some confidence. Only 9 percent has hardly any confidence.

<table>
<tr><td>➤ Data File:</td><td>HISTORY</td></tr>
<tr><td>➤ Task:</td><td>Historical Trends</td></tr>
<tr><td>➤ Variable:</td><td>20) MEDICINE</td></tr>
</table>

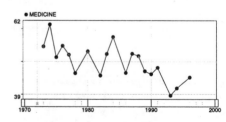

Percent having a great deal of confidence in medicine

The proportion expressing a great deal of confidence in medicine has declined a fair amount since the early 1970s. What might account for this decline?

NAME:

COURSE:

DATE:

REVIEW QUESTIONS

Based on the first part of this exercise, answer True or False to the following items:

Alcohol use is highest in the most developed nations.	T	F
Cigarette use is generally highest in Africa.	T	F
In the United States, the reported AIDS rate is highest in the Midwest.	T	F
Despite expectations, the poorer states aren't more likely than richer states to have people lacking medical insurance.	T	F
In the GSS, about half the respondents have received medical care in the past week.	T	F
In the GSS, confidence in medicine has risen overall since the 1970s.	T	F

EXPLORIT QUESTIONS

1. Should AIDS be more common in poorer nations?

    ➤ *Data File:* **NATIONS**
    ➤ *Task:* **Scatterplot**
    ➤ *Dependent Variable:* **117) AIDS**
    ➤ *Independent Variable:* **29) $ PER CAP**
    ➤ *View:* **Reg. Line**

    What conclusion would you draw from this scatterplot? Explain your answer.

2. Should infant mortality be higher in the states where more people lack health insurance?

    ➤ *Data File:* **STATES**
    ➤ *Task:* **Scatterplot**
    ➤ *Dependent Variable:* **44) CHLD MORTL**
    ➤ *Independent Variable:* **92) HEALTH.INS**
    ➤ *View:* **Reg. Line**

a. Is r statistically significant?   Yes   No

b. Do states with more people lacking health insurance tend to have higher infant mortality rates?   Yes   No

c. Do theses results suggest that lack of health insurance leads to higher infant mortality rates? Explain your answer.

3. The GSS asks whether genetic screening will do more good than harm or more harm than good.

> ➤ *Data File:* **GSS**
> ➤ *Task:* **Univariate**
> ➤ *Primary Variable:* **147) GENE GOOD**
> ➤ *View:* **Pie**

a. How would you respond to this question?   More Good

       More Harm

       Undecided

b. Why do you feel this way?

c. What percent of GSS respondents say more good?   _____%

d. Does your opinion agree with the majority of the country as indicated in the GSS?   Yes   No

4. Let's see whether income relates to views on genetic testing.

> *Data File:* **GSS**
> ➤ *Task:* **Cross-tabulation**
> ➤ *Row Variable:* **147) GENE GOOD**
> ➤ *Column Variable:* **7) FAM INCOME**
> ➤ *View:* **Tables**
> ➤ *Display:* **Column %**

a. What percent of people in the lowest income group say genetic screening
will do more good than harm? _____%

b. What percent of people in the highest income group say genetic screening
will do more good than harm? _____%

c. Is V statistically significant? Yes No

d. Is family income related to views on genetic screening? Yes No

e. Did you expect a social class difference? Why or why not?

5. In another controversial issue, the GSS asks whether people with two healthy kidneys should be
allowed to sell one of them to a hospital or organ center to be used for transplants.

> *Data File:* **GSS**
> ➤ *Task:* **Univariate**
> ➤ *Primary Variable:* **148) SELL ORGAN**
> ➤ *View:* **Pie**

a. How would you respond to this question? Not Permit

Perhaps

Permit

b. Why do you feel this way?

c. What percent of GSS respondents say selling kidneys should not
be permitted? _____%

d. Is your opinion the most common one in the country as indicated
in the GSS? Yes No

6.   Should religiosity affect views on organ selling?

> Data File: **GSS**
> > Task: **Cross-tabulation**
> > Row Variable: **148) SELL ORGAN**
> > Column Variable: **127) ATTEND**
> > View: **Tables**
> > Display: **Column %**

a.   What percent of people who attend religious services weekly feel selling
     kidneys should be permitted?   _____%

b.   What percent of people who never attend religious services feel selling
     kidneys should be permitted?   _____%

c.   Is V statistically significant?   Yes   No

d.   Is religiosity related to approval of selling kidneys?   Yes   No

e.   Did you expect a religiosity difference? Why or why not?

7.   In the preliminary section of this exercise we mapped alcohol use. Let's see whether alcohol consump-
     tion at the nation level is related to the likelihood of cirrhosis of the liver. At the same time we'll rank the
     deaths from this disease per 100,000 population and alcohol use.

> > Data File: **NATIONS**
> > Task: **Mapping**
> > Variable 1: **113) CIRRHOSIS**
> > Variable 2: **120) ALCOHOL**
> > View: **Rank**

Do these ranked lists suggest these variables are fairly similar or very different?
(Circle one.)   Fairly similar

                                                                              Very different

Compare the two ranked lists. How many of the same nations appear in
the top ten positions on both lists?   _____

8. Now let's obtain a scatterplot to reveal the association between alcohol use and cirrhosis deaths.

> Data File: **NATIONS**
> ➤ Task: **Scatterplot**
> ➤ Dependent Variable: **113) CIRRHOSIS**
> ➤ Independent Variable: **120) ALCOHOL**
> ➤ View: **Reg. Line**

In the space below, comment on what this scatterplot reveals about whether alcohol use is related to international variation in cirrhosis of the liver.

9. Let's return to AIDS worldwide.

> Data File: **NATIONS**
> ➤ Task: **Mapping**
> ➤ Variable 1: **117) AIDS**
> ➤ View: **Map**

a. In what region does AIDS appear least common? _____

b. Obtain the rankings of the nations for this variable by using the [List:Rank] or [List: Alpha] option. Then answer the following questions.

What is the rate for the United States? _____

What is Canada's rate? _____

Why do you think the U.S. rate is so much higher than Canada's?

10. Syphilis is a very serious sexually transmitted disease. Which region of the United States will have the highest number of reported cases of syphilis per 100,000 population?

> ➤ *Data File:*   **STATES**
> ➤ *Task:*   **Mapping**
> ➤ *Variable 1:*   **90) SYPHILIS**
> ➤ *View:*   **Map**

    a. Which region has the highest rate? (Circle one.)   Northeast

                                                                                                South

                                                                                               Midwest

                                                                                                 West

    b. Which region has the lowest rate? (Circle one.)   Northeast

                                                                                                South

                                                                                                 Midwest

                                                                                                 West

    c. What do you think accounts for the regional differences you just observed?

11. Locate a variable in the STATES data set that you think is related to a state's syphilis rate. Obtain a scatterplot where 90) SYPHILIS is the dependent variable and the variable you selected is the independent variable. Then answer the following questions.

    a. What independent variable did you select (name and number)?   _____

    b. Why did you feel this variable would be associated with the syphilis rate?

    c. What was the value of r in your scatterplot?   r = _____

    d. Was r statistically significant? (Circle one.)   Yes   No

    e. Was your hypothesis supported? (Circle one.)   Yes   No

◆ EXERCISE 15 ◆

POPULATION AND URBANIZATION

Tasks: Mapping, Scatterplot, Univariate, Cross-tabulation
Data Files: NATIONS, STATES, GSS

Over the last several centuries, societies around the world have grown and become increasingly urbanized. People who formerly lived in small, isolated groups have moved closer together in greater and greater numbers. This process accelerated in the nineteenth century with the advent of industrialization. As people moved closer to the new economy's place of employment, the factory, cities grew by leaps and bounds. This growth had profound consequences for almost every aspect of social life.

Demography is the study of population growth and decline. Its key concepts are *fertility*, or the average number of children born to a woman; *mortality*, or the number of deaths in a society and usually measured as the death rate, or the number of deaths per 1000 population; and *migration*, or the movement of people in and out of a specific society or location within that society. The net migration rate refers to the difference between the number of people moving in and the number moving out. Population growth or decline is a function of all three factors.

This exercise examines some key concepts and issues in the study of population and urbanization. Our primary focus will be on fertility, urbanization, and their correlates.

A DEMOGRAPHIC LOOK AROUND THE GLOBE

Demographers have long known that fertility is highest in the underdeveloped nations. Let's visualize the fertility differences they find.

➤ *Data File:* **NATIONS**
➤ *Task:* **Mapping**
➤ *Variable 1:* **9) FERTILITY**
➤ *View:* **Map**

FERTILITY – NUMBER OF CHILDREN BORN TO AVERAGE WOMAN IN HER LIFETIME
(PON, 1996)

This map provides one of the clearest distinctions between developed and underdeveloped nations that we've seen in this workbook. Fertility is generally highest in Africa and some Asian nations and lowest in North America and Europe.

Why do these fertility patterns exist? Demographers cite several reasons that we'll test separately. First, and in no particular order, women in underdeveloped nations are often prized above all for their

215

childbearing ability. Where women are so prized, fertility should be higher. Let's test both parts of this hypothesis before turning to the other reasons.

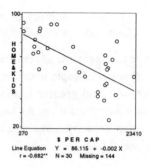

Data File:	**NATIONS**
➤ Task:	**Scatterplot**
➤ Dependent Variable:	**80) HOME&KIDS**
➤ Independent Variable:	**29) $ PER CAP**
➤ View:	**Reg. Line**

So far, so good. The less developed the nation, the more likely its citizens believe that a woman wants a home and children above all (r = –.68**). Let's now look at the relationship between the home and children variable and the actual fertility rate in nations.

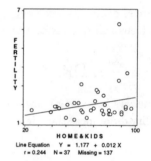

Data File:	**NATIONS**
Task:	**Scatterplot**
➤ Dependent Variable:	**9) FERTILITY**
➤ Independent Variable:	**80) HOME&KIDS**
➤ View:	**Reg. Line**

Not so good. The relationship between acceptance of women's traditional home role and fertility is in the expected direction but falls short of statistical significance (r = .24). One problem with using the independent variable chosen here is that it's from the World Values Study (WVS), which, although it includes most of the world's population in its 40-odd nations, doesn't include most of the underdeveloped nations, which typically have very small populations. Thus most WVS nations have relatively low fertility to begin with. Perhaps we'll have more success in illustrating the other reasons demographers cite for the high fertility of underdeveloped nations.

A second such reason is that contraception is uncommon in underdeveloped nations. When contraception is used less, fertility should obviously be higher.

Discovering Sociology

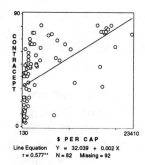

Data File: **NATIONS**
Task: **Scatterplot**
➤ *Dependent Variable:* **13) CONTRACEPT**
➤ *Independent Variable:* **29) $ PER CAP**
➤ *View:* **Reg. Line**

The less developed the nation, the lower its contraception use (r = .58**). Let's look at fertility rates and the use of contraception.

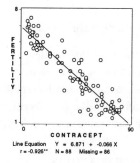

Data File: **NATIONS**
Task: **Scatterplot**
➤ *Dependent Variable:* **9) FERTILITY**
➤ *Independent Variable:* **13) CONTRACEPT**
➤ *View:* **Reg. Line**

This is perhaps the strongest relationship we've seen in the workbook thus far (r = −.93**): the less the contraception use, the higher the fertility.

A third reason for high fertility in underdeveloped nations has to do with their national economies. Because the poorest nations are primarily agricultural, children are needed for economic reasons to help the family do its agricultural work. Let's see if the agricultural nations have higher fertility.

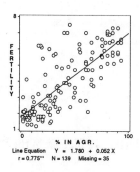

Data File: **NATIONS**
Task: **Scatterplot**
Dependent Variable: **9) FERTILITY**
➤ *Independent Variable:* **33) % IN AGRI.**
➤ *View:* **Reg. Line**

The agricultural nations are indeed much more likely to have higher fertility (r = .78**).

A final reason has to do with childhood mortality. In societies with high childhood mortality, families cannot assume that all their children will reach adulthood, and having greater numbers of children increases the chances of an adequate number's ultimately surviving. If this is true, high childhood mortality should be linked to higher fertility.

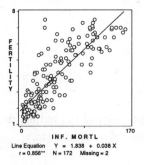

Data File: **NATIONS**
Task: **Scatterplot**
Dependent Variable: **9) FERTILITY**
➤ Independent Variable: **11) INF. MORTL**
➤ View: **Reg. Line**

Line Equation   Y = 1.838 + 0.038 X
r = 0.856**   N = 172   Missing = 2

Nations with high childhood mortality are much more likely to have higher fertility (r = .86**).

None of the correlations just examined prove that the demographers' reasons for high fertility in underdeveloped nations are correct. In some of the correlations, causal order remains a question, and in some it's possible that the relationship is spurious if some third factor affects both the independent variable and fertility. But overall the correlations are consistent with the reasons demographers cite for the underdeveloped world–high fertility connection.

POPULATION AND URBANIZATION IN THE UNITED STATES

One of the most important population issues in the United States today is teenage pregnancies and births. To explore this issue, let's begin with the percent of births in each state to women under the age of 20.

➤ Data File: **STATES**
➤ Task: **Mapping**
➤ Variable 1: **43) MA<AGE20**
➤ View: **Map**

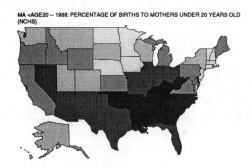

MA <AGE20 -- 1988: PERCENTAGE OF BIRTHS TO MOTHERS UNDER 20 YEARS OLD (NCHS)

Teenage births are highest in the South.

What explains this geographic pattern? You've probably heard that most teenage mothers come from poor, uneducated families. Let's see whether the poorest and least-educated states have higher rates of teenage births.

Data File: **STATES**
➤ Task: **Scatterplot**
➤ Dependent Variable: **43) MA<AGE20**
➤ Independent Variable: **55) %POOR**
➤ View: **Reg. Line**

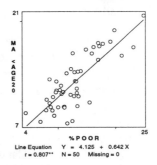

The poorer the state, the higher its teenage birth rate (r = .81**). Let's see what the relationship is to education.

Data File: **STATES**
Task: **Scatterplot**
Dependent Variable: **43) MA<AGE20**
➤ Independent Variable: **76) DROPOUTS**
➤ View: **Reg. Line**

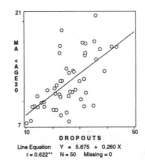

States with higher dropout rates have higher teenage birth rates (r = .62**).

Time for the GSS. One GSS item asks respondents how many children they've ever had. Let's see what women report. (We'll use the subset option to limit the analysis to women.)

➤ Data File: **GSS**
➤ Task: **Univariate**
➤ Primary Variable: **56) # CHILDREN**
➤ Subset Variable: **45) GENDER**
➤ Subset Categories: **Include: 1) Female**
➤ View: **Pie**

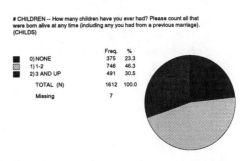

The option for selecting a subset variable is located on the same screen you use to select other variables. For this example, select 45) GENDER as the subset variable. A window will appear that shows you the categories of the subset variable. Select 1) Female as your subset category and choose the [Include] option. Then click [OK] and continue as usual.

Almost 31 percent of the women respondents have had three or more children, and another 46.3 percent have had 1–2 children. The remainder, 23.3 percent, have had no children.

What factors explain the number of children a woman has? We saw earlier that fertility was higher in poorer nations. Will GSS women whose parents had lower incomes be more likely than respondents with wealthier parents to have three or more children? Our independent (column) variable will be parents' occupational prestige.

Data File: **GSS**
➤ Task: **Cross-tabulation**
➤ Row Variable: **56) # CHILDREN**
➤ Column Variable: **19) PARS.PRESG**
➤ Subset Variable: **45) GENDER**
➤ Subset Categories: **Include: 1) Female**
➤ View: **Tables**
➤ Display: **Column %**

19) PARS.PRESG

56) # CHILDREN	BOTH LOW	BOTH MEDI	BOTH HIGH
NONE	19.5%	26.3%	42.9%
1-2	50.8%	47.4%	41.1%
3 AND UP	29.7%	26.3%	16.1%
TOTAL	100.0%	100.0%	100.0%

V=0.166**

Note that you again need to select GENDER as a subset variable.

Respondents whose parents both held low-prestige jobs are almost twice as likely as those whose parents both held high-prestige jobs to have three or more children (V = .17**).

Now let's consider the woman's own schooling and hypothesize that women with less education will have more children.

Data File: **GSS**
Task: **Cross-tabulation**
Row Variable: **56) # CHILDREN**
➤ Column Variable: **6) EDUCATION**
➤ Subset Variable: **45) GENDER**
➤ Subset Categories: **Include: 1) Female**
➤ View: **Tables**
➤ Display: **Column %**

6) EDUCATION

56) # CHILDREN	NO HS GRA	HS GRAD	SOME COL	COLL GRA
NONE	13.2%	15.5%	24.1%	38.4%
1-2	40.1%	50.9%	46.4%	45.5%
3 AND UP	46.7%	33.6%	29.4%	16.1%
TOTAL	100.0%	100.0%	100.0%	100.0%

V=0.196**

Women who have graduated from college are almost three times less likely than those without a high school degree to have three or more children, and substantially more likely to have no kids (V = .20**). Notice that causal order here is unclear, because it's possible that having children restricted a woman's ability to go to, and graduate from, college. Still, most demographers and other scholars feel that education does help explain how many children a woman has.

What about religious preference? Do you think Catholic women have more children than Protestant or Jewish women?

<table>
<tr><td>Data File:</td><td>GSS</td></tr>
<tr><td>Task:</td><td>Cross-tabulation</td></tr>
<tr><td>Row Variable:</td><td>56) # CHILDREN</td></tr>
<tr><td>➤ Column Variable:</td><td>126) RELIGION</td></tr>
<tr><td>➤ Subset Variable:</td><td>45) GENDER</td></tr>
<tr><td>➤ Subset Categories:</td><td>Include: 1) Female</td></tr>
<tr><td>➤ View:</td><td>Tables</td></tr>
<tr><td>➤ Display:</td><td>Column %</td></tr>
</table>

126) RELIGION

56) # CHILDREN	PROTESTA	CATHOLIC	JEWISH
NONE	19.3%	26.4%	24.4%
1-2	47.7%	41.2%	53.3%
3 AND UP	33.0%	32.5%	22.2%
TOTAL	100.0%	100.0%	100.0%

V=0.062*

Catholics do *not* have more children than Protestants; the major difference in the table is that Jewish women have fewer children than Protestant and Catholic women (V = .06*).

Now let's turn our attention to people living in cities and in rural areas. We'll look at medium- and large-city residents who have lived in the same city all their lives and compare them to rural residents who have lived in the same rural town or farm all their lives.

Going back to at least Ferdinand Toennies a century ago, sociologists have depicted cities as cold, alienating places with weak social bonds and high amounts of distrust. Other sociologists have criticized this view as an unfair stereotype and point to closely knit neighborhoods within large cities. Let's see which view, if either, GSS data support.

The GSS asks respondents how close they feel to their neighborhood or village. If the critics of cities are right, urban respondents should feel less close than rural residents to their immediate locations.

<table>
<tr><td>Data File:</td><td>GSS</td></tr>
<tr><td>Task:</td><td>Cross-tabulation</td></tr>
<tr><td>➤ Row Variable:</td><td>84) CLSENEI</td></tr>
<tr><td>➤ Column Variable:</td><td>94) URBAN?</td></tr>
<tr><td>➤ View:</td><td>Tables</td></tr>
<tr><td>➤ Display:</td><td>Column %</td></tr>
</table>

94) URBAN?

84) CLSENEI	URBAN	RURAL
CLOSE	56.5%	71.8%
NOT CLOSE	43.5%	28.2%
TOTAL	100.0%	100.0%

V=0.150**

Urban residents are less likely (56.5 percent) than rural residents (71.8 percent) to feel close to their immediate locations (V = .15**). This table supports the city critics.

Another variable asks how close they feel to their town or city.

<table>
<tr><td>Data File:</td><td>GSS</td></tr>
<tr><td>Task:</td><td>Cross-tabulation</td></tr>
<tr><td>➤ Row Variable:</td><td>85) CLSETOWN</td></tr>
<tr><td>➤ Column Variable:</td><td>94) URBAN?</td></tr>
<tr><td>➤ View:</td><td>Tables</td></tr>
<tr><td>➤ Display:</td><td>Column %</td></tr>
</table>

85) CLSETOWN	94) URBAN?	
	URBAN	RURAL
CLOSE	71.7%	75.7%
NOT CLOSE	28.3%	24.3%
TOTAL	100.0%	100.0%

V=0.042

Taking into account statistical significance, urban and rural residents feel equally close to their cities or towns (V = .04). So are the critics of cities correct or not?

Critics of cities would predict that urban residents are less trusting of others than rural residents. Let's check this out by using responses to the question "Generally speaking, would you say that most people can be trusted or that you can't be too careful in dealing with people?"

<table>
<tr><td>Data File:</td><td>GSS</td></tr>
<tr><td>Task:</td><td>Cross-tabulation</td></tr>
<tr><td>➤ Row Variable:</td><td>137) TRUSTED</td></tr>
<tr><td>➤ Column Variable:</td><td>94) URBAN?</td></tr>
<tr><td>➤ View:</td><td>Tables</td></tr>
<tr><td>➤ Display:</td><td>Column %</td></tr>
</table>

137) TRUSTED	94) URBAN?	
	URBAN	RURAL
CAN TRUST	31.5%	29.8%
BE CAREFUL	68.5%	70.2%
TOTAL	100.0%	100.0%

V=0.017

Urban and rural residents are equally distrustful (V = .02).

The GSS also asks, "Do you think most people would try to take advantage of you if they got a chance, or would they try to be fair?"

<table>
<tr><td>Data File:</td><td>GSS</td></tr>
<tr><td>Task:</td><td>Cross-tabulation</td></tr>
<tr><td>➤ Row Variable:</td><td>136) ADVANTAGE?</td></tr>
<tr><td>➤ Column Variable:</td><td>94) URBAN?</td></tr>
<tr><td>➤ View:</td><td>Tables</td></tr>
<tr><td>➤ Display:</td><td>Column %</td></tr>
</table>

136) ADVANTAGE?	94) URBAN?	
	URBAN	RURAL
TAKE ADVAN	50.4%	49.4%
BE FAIR	49.6%	50.6%
TOTAL	100.0%	100.0%

V=0.009

Both sets of residents are also equally likely to think that people take advantage (V = .01).

None of these last three tables supports the bleak view of the critics of cities.

NAME:

COURSE:

DATE:

REVIEW QUESTIONS

Based on the first part of this exercise, answer True or False to the following items:

The more agricultural nations have higher fertility.	T	F
Fertility tends to be very high in Africa.	T	F
In the United States, teenage births are highest in the West.	T	F
In the United States, the poorer a state, the higher its percent of births to teenage mothers.	T	F
In the GSS, Catholic women have had more children than Protestant women.	T	F
GSS data support the common view that cities are cold, alienating places.	T	F

EXPLORIT QUESTIONS

1. At the international level, is fertility related to believing that the ideal family size is three or more children?

> *Data File:* **NATIONS**
> *Task:* **Scatterplot**
> *Dependent Variable:* **9) FERTILITY**
> *Independent Variable:* **10) LARGE FAML**
> *Display:* **Reg. Line**

a. Summarize the relationship depicted by this scatterplot.

b. As you know, a scatterplot doesn't necessarily indicate causal order. In this one, does it make sense to think that a belief in large families affects fertility, or does it make sense to think that fertility affects beliefs about family size? Explain your answer.

2. In the United States, do the more urban states have higher suicide rates?

> ➤ *Data File:* **STATES**
> ➤ *Task:* **Scatterplot**
> ➤ *Dependent Variable:* **34) SUICIDE**
> ➤ *Independent Variable:* **11) %URBAN**
> ➤ *Display:* **Reg. Line**

   a. Is r statistically significant?   Yes   No

   b. Are suicide rates higher in the more urbanized states?   Yes   No

   c. Would Durkheim have predicted that suicide rates should be higher in more urbanized areas? Explain your answer.

3. An important variable in population studies is the age at which a woman has her first child. What factors might account for this age?

> ➤ *Data File:* **GSS**
> ➤ *Task:* **Cross-tabulation**
> ➤ *Row Variable:* **58) AGE KD BORN**
> ➤ *Column Variable:* **19) PARS.PRESG**
> ➤ *Subset Variable:* **45) GENDER**
> ➤ *Subset Categories:* **Include: 1) Female**
> ➤ *View:* **Tables**
> ➤ *Display:* **Column %**

   > The option for selecting a subset variable is located on the same screen you use to select other variables. For this example, select 45) GENDER as the subset variable. A window will appear that shows you the categories of the subset variable. Select 1) Female as your subset category and choose the [Include] option. Then click [OK] and continue as usual.

   a. What percent of female respondents with parents with high-prestige jobs had their first child before age 20?   _____%

   b. What percent of female respondents with parents with low-prestige jobs had their first child before age 20?   _____%

   c. Is V statistically significant?   Yes   No

   d. Is parents' social class related to the age at which women have their
first child?   Yes   No

4.   Now let's see whether urban or rural residence affects the age at which a woman has her first child.

> Data File: **GSS**
> Task: **Cross-tabulation**
> Row Variable: **58) AGE KD BORN**
> ➤ Column Variable: **94) URBAN?**
> ➤ Subset Variable: **45) GENDER**
> ➤ Subset Categories: **Include: 1) Female**
> ➤ View: **Tables**
> ➤ Display: **Column %**

   a. What percent of urban, female respondents had children before age 20?   _____%

   b. What percent of rural, female respondents had children before age 20?   _____%

   c. Is V statistically significant?   Yes   No

   d. Is urban/rural residence related to the age at which women have
their first child?   Yes   No

5.   It is often thought that urban residents are more tolerant of controversial behaviors than rural residents. Let's test this assumption.

> Data File: **GSS**
> Task: **Cross-tabulation**
> ➤ Row Variable: **74) HOMO.SEX**
> ➤ Column Variable: **94) URBAN?**
> ➤ View: **Tables**
> ➤ Display: **Column %**

> **Notice, if you are continuing from the previous question, that you must exit the task, delete the subset variable, or clear all variables to remove the subset variable.**

   a. What percent of urban residents think homosexuality is always wrong?   _____%

   b. What percent of rural residents think homosexuality is always wrong?   _____%

    c.   Is V statistically significant?   Yes   No

    d.   Are urban residents more tolerant of homosexuality?   Yes   No

6.   Now let's examine views on premarital sex.

>   Data File: **GSS**
>   Task: **Cross-tabulation**
> ➤ Row Variable: **72) PREM.SEX**
> ➤ Column Variable: **94) URBAN?**
>   ➤ View: **Tables**
>   ➤ Display: **Column %**

    a.   What percent of urban residents think premarital sex is always wrong?   _____%

    b.   What percent of rural residents think premarital sex is always wrong?   _____%

    c.   Is V statistically significant?   Yes   No

    d.   Are urban residents more tolerant of premarital sex?   Yes   No

7.   Based on your answers to Questions 5 and 6, would you say that urban residents are more tolerant than rural residents of controversial behaviors? What differences between urban and rural areas might explain your answer?

8.   Some scholars believe that to lower fertility we must first give women more power and freedom to control their lives. In this way of thinking, when women have options outside the home and can decide when they want to marry and have children, they will decide to have fewer babies and, con-comitantly, not be forced to conceive babies they don't want. If this is true, nations where women are the most empowered in these and other ways should have lower fertility. Let's test this hypothesis.

> ➤   Data File: **NATIONS**
> ➤   Task: **Scatterplot**
> ➤ Dependent Variable: **9) FERTILITY**
> ➤ Independent Variable: **76) FEM POWER**
> ➤   Display: **Reg. Line**

a. Taking into account r and its statistical significance, do the nations where women are most empowered have lower fertility? (Circle one) Yes No

b. Could the relationship depicted in the scatterplot be spurious? For example, could the more educated nations have more female empowerment and also lower fertility rates? Or do you think that female empowerment really does lower fertility? Explain your answer.

9. In Questions 5–7 we examined whether urban residents in the GSS were more tolerant than rural residents of certain controversial behaviors. Let's explore the relationship between urbanism and tolerance with two behaviors included in the NATIONS data set.

> Data File: **NATIONS**
> Task: **Scatterplot**
> ➤ Dependent Variable: **112) PROSTITUTE**
> ➤ Independent Variable: **4) URBAN %**
> ➤ Display: **Reg. Line**

> Data File: **NATIONS**
> Task: **Scatterplot**
> ➤ Dependent Variable: **111) GAY SEX**
> ➤ Independent Variable: **4) URBAN %**
> ➤ Display: **Reg. Line**

Taking into account the results depicted in both scatterplots, are the more urban nations more tolerant of controversial behaviors than the less urban nations? Explain your answer.

10. Based on our knowledge of teenage mothers, which part of the United States should have the highest birth rate?

> ➤ Data File: **STATES**
> ➤ Task: **Mapping**
> ➤ Variable 1: **41) BIRTHS**
> ➤ View: **Map**

a. Generally speaking, which region *has* the highest birth rate? _____

b. Why do you think this region has the highest rate?

11. Do religious women have more children than less religious women?

> ➤ Data File: **GSS**
> ➤ Task: **Cross-tabulation**
> ➤ Row Variable: **56) # CHILDREN**
> ➤ Column Variable: **128) PRAY**
> ➤ Subset Variable: **45) GENDER**
> ➤ Subset Category: **Include: 1) Female**
> ➤ View: **Tables**
> ➤ Display: **Column %**

a. What percent of women who pray daily have three or more children?   _____%

b. What percent of women who pray less than weekly have three or
more children?   _____%

c. It's possible that this relationship is spurious—that some third factor affects both religiosity and
the number of children a woman has. What would be one possible such factor? Explain your
answer.

SOCIAL CHANGE AND MODERNIZATION

Tasks: Scatterplot, Mapping, Historical Trends, Univariate, Cross-tabulation
Data Files: NATIONS, STATES, HISTORY, GSS

Much of this workbook has been about social change. Time after time we've seen how developed, industrial societies differ from the underdeveloped, agricultural ones. In doing so, we've seen both the good and bad effects of modernization. In looking at the United States, we've seen that some attitudes and behaviors have changed in the last 25 years—admittedly a short time span—but that others have not. Many scholars trace the changes that have occurred to the turbulent 1960s, when the civil rights movement in the South, the Vietnam antiwar movement, and other social-change efforts altered the course of our nation.

This exercise summarizes some of the major changes industrialization has brought around the world and reviews some of the major ways in which the United States has changed since the 1960s. Several worksheet questions focus on views of, and involvement in, political protest, one type of the collective action that people engage in to bring about or prevent social change.

SOCIAL CHANGE AND MODERNIZATION IN GLOBAL PERSPECTIVE

Modernization, as we've seen, has had important consequences for the life chances, behaviors, and attitudes of societies around the world. Let's review some of the changes, both good and bad, that modernization has brought. Our independent variable will be the percent of a nation's gross domestic product *not* accounted for by agriculture; this is an admittedly very rough measure of the degree to which a society is industrialized and thus modern. We'll use this variable to see what difference modernization makes in four broad areas of belief and practice.

Modernization and Life Chances

➤ *Data File:* **NATIONS**
➤ *Task:* **Scatterplot**
➤ *Dependent Variable:* **40) EDUCATION**
➤ *Independent Variable:* **143) % INDUS $**
➤ *View:* **Reg. Line**

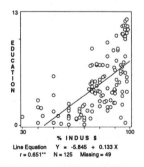

Line Equation   Y = -5.845 + 0.133 X
r = 0.651**   N = 125   Missing = 49

Modern societies have much higher levels of education (r = .65**).

Data File: **NATIONS**
Task: **Scatterplot**
➤ *Dependent Variable:* **11) INF.MORTL**
➤ *Independent Variable:* **143) % INDUS $**
➤ *View:* **Reg. Line**

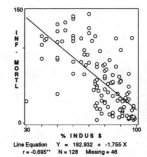

Modern societies also have much lower rates of infant mortality (r = −.70**) . . .

Data File: **NATIONS**
Task: **Scatterplot**
➤ *Dependent Variable:* **12) MOM MORTAL**
➤ *Independent Variable:* **143) % INDUS $**
➤ *View:* **Reg. Line**

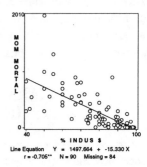

. . . much lower rates of death during childbirth (r = −.71**) . . .

Data File: **NATIONS**
Task: **Scatterplot**
➤ *Dependent Variable:* **18) DEATH RATE**
➤ *Independent Variable:* **143) % INDUS $**
➤ *View:* **Reg. Line**

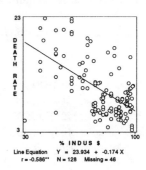

. . . and much lower rates of death from all causes (r = −.59**).

But modernization also has its personal costs.

Discovering Sociology

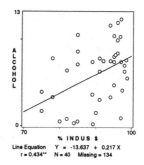

Data File: **NATIONS**
Task: **Scatterplot**
➤ Dependent Variable: **120) ALCOHOL**
➤ Independent Variable: **143) % INDUS $**
➤ View: **Reg. Line**

People in the most industrialized nations are apt to drink more alcohol (r = .43**) . . .

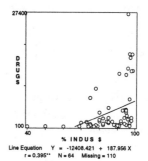

Data File: **NATIONS**
Task: **Scatterplot**
➤ Dependent Variable: **118) DRUGS**
➤ Independent Variable: **143) % INDUS $**
➤ View: **Reg. Line**

. . . to use narcotic drugs (r = .40**) . . .

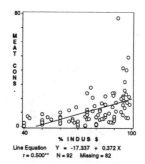

Data File: **NATIONS**
Task: **Scatterplot**
➤ Dependent Variable: **24) MEAT CONS.**
➤ Independent Variable: **143) % INDUS $**
➤ View: **Reg. Line**

. . . to eat (high-fat!) meat (r = .50**) . . .

Data File: **NATIONS**
Task: **Scatterplot**
➤ *Dependent Variable:* **124) CIGARETTES**
➤ *Independent Variable:* **143) % INDUS $**
➤ *View:* **Reg. Line**

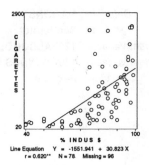

. . . and to smoke cigarettes (r = .62**).

Modernization, Technology, and the Environment

Not surprisingly, modern nations are especially likely to enjoy the benefits of modern technology, but those benefits, too, have a cost, in this case a global one.

Data File: **NATIONS**
Task: **Scatterplot**
➤ *Dependent Variable:* **35) TV 1000**
➤ *Independent Variable:* **143) % INDUS $**
➤ *View:* **Reg. Line**

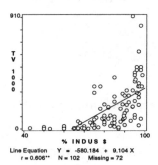

On the upside, modern societies are much more likely to have such things as televisions (r = .61**) . . .

Data File: **NATIONS**
Task: **Scatterplot**
➤ *Dependent Variable:* **34) CARS/1000**
➤ *Independent Variable:* **143) % INDUS $**
➤ *View:* **Reg. Line**

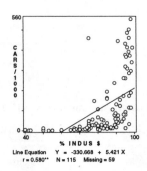

. . . and cars (r = .58**).

But all that technology has an environmental cost.

Data File: **NATIONS**
Task: **Scatterplot**
➤ Dependent Variable: **31) ELECTRIC**
➤ Independent Variable: **143) % INDUS $**
➤ View: **Reg. Line**

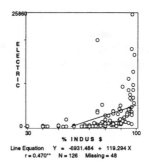

Modern societies use much more electricity than traditional ones (r = .47**) . . .

Data File: **NATIONS**
Task: **Scatterplot**
➤ Dependent Variable: **26) GREENHOUSE**
➤ Independent Variable: **143) % INDUS $**
➤ View: **Reg. Line**
➤ Find: **Outlier/Remove**

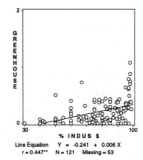

Find the outlier by selecting the [Outlier] option. A box will appear around the dot representing the outlier case. Remove this case by clicking the [Remove] button. Notice the change in the r value after removing the outlier case.

. . . and contribute more to global emissions (r = .45**) and thus to a possible greenhouse effect.

Modernization and Gender, Racial, and Ethnic Inequality

Modernization brings with it not only better life chances overall and improved technology, but also greater gender equality and less racial and ethnic prejudice.

Data File: **NATIONS**
Task: **Scatterplot**
➤ Dependent Variable: **75) GENDER EQ**
➤ Independent Variable: **143) % INDUS $**
➤ View: **Reg. Line**

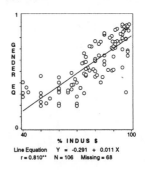

Modern societies have much more gender equality than less modern societies (r = .81**).

Data File: **NATIONS**
Task: **Scatterplot**
➤ Dependent Variable: **80) HOME&KIDS**
➤ Independent Variable: **143) % INDUS $**
➤ View: **Reg. Line**

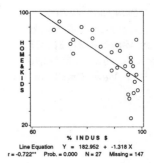

In line with this fact, people in the more modern nations are much less likely to feel a woman's place is in the home (r = −.72**).

The NATIONS data set doesn't have measures of racial equality, but does include measures of racial and ethnic prejudice.

Data File: **NATIONS**
Task: **Scatterplot**
➤ Dependent Variable: **61) RACISM**
➤ Independent Variable: **143) % INDUS $**
➤ View: **Reg. Line**

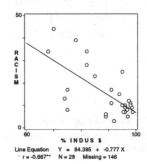

Looking at one such measure, people in the more modern nations are much less likely to say they wouldn't want members of other races as their neighbors (r = −.67**). Racial prejudice declines as societies become more modern.

Modernization and Tolerance for Controversial Behaviors

The last area in which we'll examine modernization's effects is tolerance. One hallmark of modern societies is greater tolerance of some of the behaviors that less modern societies often condemn. Although disapproval of such behaviors may still be common in modern societies, it's less common than in the less modern ones.

Data File: **NATIONS**
Task: **Scatterplot**
➤ Dependent Variable: **111) GAY SEX**
➤ Independent Variable: **143) % INDUS $**
➤ View: **Reg. Line**

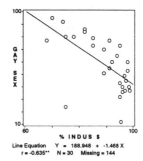

People in the more modern societies are less apt to think that homosexuality is never acceptable (r = −.64**).

Data File: **NATIONS**
Task: **Scatterplot**
➤ Dependent Variable: **112) PROSTITUTE**
➤ Independent Variable: **143) % INDUS $**
➤ View: **Reg. Line**

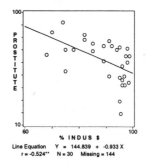

They're also less likely to think that prostitution is never acceptable (r = −.52**).

SOCIAL CHANGE IN THE UNITED STATES

We turn now to the GSS to explore a few ways in which Americans have changed their attitudes and behaviors in the last quarter century, and a few areas in which they've not changed. This period of our nation's history followed on the heels of the tumultuous 1960s and thus represents a fascinating time in which to explore whether the '60s' movements and other events have had an enduring impact. The last quarter century has also seen significant political developments and controversies that may have changed people's views or practices.

Sex and Drugs

We'll begin our exploration with a look at views and practices having to do with sex and drugs.

➤ *Data File:* **HISTORY**
 ➤ *Task:* **Historical Trends**
➤ *Variable:* **16) HOMO.SEX**

Percent saying homosexual sex is always wrong

The percent saying homosexuality is "always wrong" stayed fairly stable from the early 1970s through the early 1990s, but has declined since that time.

Data File: **HISTORY**
 Task: **Historical Trends**
➤ *Variable:* **15) PREM.SEX**

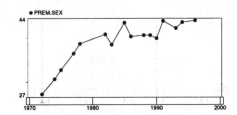

Percent saying premarital sex is not wrong

The percent saying premarital sex is "not wrong" rose through the 1970s and early 1980s and remains higher today than 25 years ago.

Data File: **HISTORY**
 Task: **Historical Trends**
➤ *Variable:* **14) XMAR.SEX**

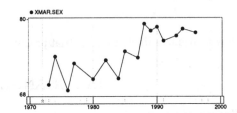

Percent saying extramarital sex is always wrong

Although views on homosexuality and premarital sex have become more lenient in the past 25 years, views on extramarital sex—adultery—have become more intolerant.

Data File: **HISTORY**
 Task: **Historical Trends**
➤ *Variable:* **24) ABORT ANY**

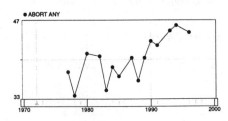

Percent favoring legal abortion for any reason

Support for legalized abortion for any reason has risen in the past 10 years.

Discovering Sociology

Data File: **HISTORY**
Task: **Historical Trends**
➤ Variable: **3) GRASS?**

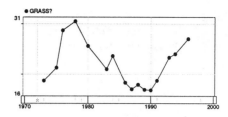

Percent saying marijuana should be made legal

The percent thinking marijuana should be made legal rose rapidly from the early 1970s through the late 1970s and then declined, only to rise again in the 1990s.

Data File: **HISTORY**
Task: **Historical Trends**
➤ Variable: **25) SMOKE?**

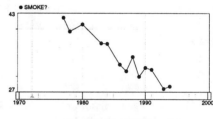

Percent smoking cigarettes

Cigarette smoking has dropped considerably in the past 25 years.

Data File: **HISTORY**
Task: **Historical Trends**
➤ Variable: **26) DRINK?**

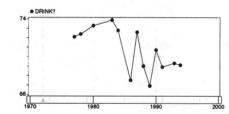

Percent drinking alcohol instead of abstaining

Drinking, however, has declined only slightly.

Political Beliefs

The second area we examine concerns political beliefs. We'll start with the percent who agree that "most public officials are not really interested in the problems of the average man."

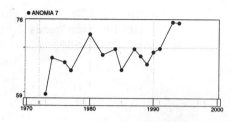

Data File: **HISTORY**
Task: **Historical Trends**
➤ Variable: **27) ANOMIA 7**

Percent thinking politicians don't care about what the average person thinks

This percent has fluctuated since the early 1970s. It rose in the latter half of that decade and declined during Ronald Reagan's presidency, then rose again. It remains higher now than at the beginning of this period.

Has the United States become more conservative during the last quarter century? To find out, let's see whether today's GSS respondents view themselves as more conservative than their counterparts 25 years ago.

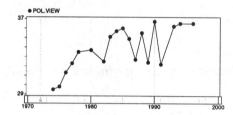

Data File: **HISTORY**
Task: **Historical Trends**
➤ Variable: **17) POL.VIEW**

Percent saying their views are conservative

The country has become somewhat more conservative in the past 25 years.

Gender and Racial Issues

Our third area is gender and racial issues.

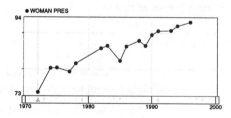

Data File: **HISTORY**
Task: **Historical Trends**
➤ Variable: **30) WOMAN PRES**

Percent willing to vote for a qualified woman for president

The percent of the public saying it would vote for a qualified woman for president has risen substantially since the early 1970s.

Discovering Sociology

Data File: **HISTORY**
Task: **Historical Trends**
➤ Variable: **33) WOMEN WORK**

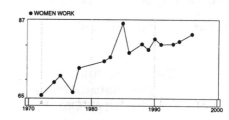

Percent approving of a married woman working outside of the home

So has the percent approving of a married woman working outside the home.

Data File: **HISTORY**
Task: **Historical Trends**
➤ Variable: **6) BLACK PRES**

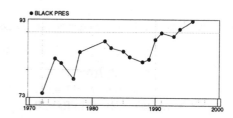

Percent willing to vote for a qualified African American for president

Despite some fluctuations, the percent of the public saying it would vote for a qualified African American for president has risen substantially since the early 1970s.

Data File: **HISTORY**
Task: **Historical Trends**
➤ Variable: **7) RACE SEG**

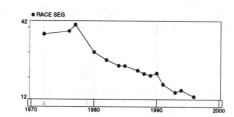

Percent favoring racial segregation in housing

Meanwhile, agreement that white people have a right to keep African Americans out of their neighborhoods has dropped substantially.

Data File: **HISTORY**
Task: **Historical Trends**
➤ Variable: **5) INTERMAR?**

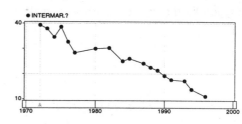

Percent agreeing with laws against racial intermarriage

As has the percent supporting laws against racial intermarriage.

Religion and Religious Issues

Our last area concerns religion.

Data File: **HISTORY**
    Task: **Historical Trends**
➤ Variable: **31) ATTEND**

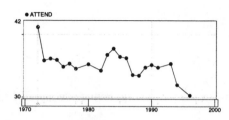

Percent attending religious services at least once per week

Weekly religious attendance has generally declined since the early 1970s.

Data File: **HISTORY**
    Task: **Historical Trends**
➤ Variable: **32) SCH.PRAYER**

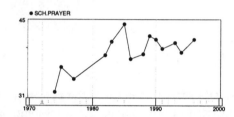

Percent approving Supreme Court decision banning prayer in public schools

Support for the U.S. Supreme Court's prohibition of prayer in the public schools has risen slightly during this period.

Data File: **HISTORY**
    Task: **Historical Trends**
➤ Variable: **35) ATHEIST SP**

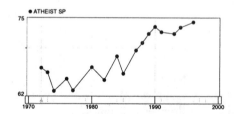

Percent approving of the right of an atheist to make a speech

Support for the right of an atheist to make a speech has also risen.

WORKSHEET

NAME:

COURSE:

DATE:

EXERCISE
16

REVIEW QUESTIONS

Based on the first part of this exercise, answer True or False to the following items:

The more modern a nation, the more likely its residents are to eat meat.	T	F
The more modern a nation, the greater its gender equality.	T	F
During the 1980s, U.S. disapproval of homosexuality dropped substantially.	T	F
Judging from GSS data in the HISTORY file, Americans have become more racially prejudiced since the 1970s.	T	F
Judging from GSS data in the HISTORY file, Americans have become less sexist since the 1970s.	T	F
Religious attendance has increased since the early 1970s.	T	F

EXPLORIT QUESTIONS

1. The brief exploration earlier of changes in several kinds of beliefs and practices during the last quarter century indicates that Americans have, overall, slightly abandoned traditional beliefs and practices on matters such as sexuality and drugs, gender, race, and religion. Do you think the changes we've seen in this section have been good or bad for the United States? Why do you feel the way you do?

2. The social movements in the 1960s engaged in various kinds of "collective action" to achieve their aims. Although the 1960s are long gone, some segments of the population continue to engage in such activity. First we'll see how many attend protest meetings.

> ➤ *Data File:* **GSS**
> ➤ *Task:* **Univariate**
> ➤ *Primary Variable:* **115) R PROTEST3**
> ➤ *View:* **Pie**

What percent of the sample had never attended a protest meeting
in the past 5 years?   _____%

3. Now we'll examine involvement in a protest march or demonstration.

             Data File: **GSS**
                  Task: **Univariate**
    ➤ Primary Variable: **116) R PROTEST4**
              ➤ View: **Pie**

        What percent of the sample had never gone to a protest march or
        demonstration in the past 5 years?   _____%

4. Does education predict protest meeting attendance?

             Data File: **GSS**
               ➤ Task: **Cross-tabulation**
       ➤ Row Variable: **115) R PROTEST3**
    ➤ Column Variable: **6) EDUCATION**
              ➤ View: **Tables**
           ➤ Display: **Column %**

    a. What percent of people with a college degree had never gone to a protest
        meeting in the prior 5 years?   _____%

    b. What percent of people without a high school degree had never gone to a
        protest meeting in the prior 5 years?   _____%

    c. Is V statistically significant?   Yes   No

    d. In your own words, summarize the major finding of this table.

5. One type of collective action, civil disobedience, involves breaking the law for reasons of conscience.

             Data File: **GSS**
               ➤ Task: **Univariate**
    ➤ Primary Variable: **124) OBEY LAW**
              ➤ View: **Pie**

What percent think people should obey their consciences even if it means breaking the law? _____%

6. Does education predict support for civil disobedience?

> *Data File:* **GSS**
> ➤ *Task:* **Cross-tabulation**
> ➤ *Row Variable:* **124) OBEY LAW**
> ➤ *Column Variable:* **6) EDUCATION**
> ➤ *View:* **Tables**
> ➤ *Display:* **Column %**

a. What percent of people without high school degrees feel people should follow their consciences? _____%

b. What percent of people with college degrees feel people should follow their consciences? _____%

c. Is V statistically significant?   Yes   No

d. What sociological reasons might account for the major finding in this table?

e. Do you feel that people should follow their consciences even if it means breaking the law? Explain your answer.

7. The NATIONS data set includes a measure of the percent who feel "the entire way our society is organized must be radically changed by revolutionary action."

> ➤ *Data File:* **NATIONS**
> ➤ *Task:* **Mapping**
> ➤ *Variable 1:* **82) REVOLUTION**
> ➤ *View:* **Rank**

a. Which part of the world is most likely to think revolution is necessary? _____

b. Which part of the world is least likely to think revolution is necessary? _____

c. What would be one explanation that helps account for this geographic difference?

8. If, as we saw in the preliminary section, the industrialized nations drink more, are their citizens also more likely to die from cirrhosis of the liver?

> Data File: **NATIONS**
> ➤ Task: **Scatterplot**
> ➤ Dependent Variable: **113) CIRRHOSIS**
> ➤ Independent Variable: **143) % INDUS $**
> ➤ Display: **Reg. Line**

a. What is the value of r?   r = _____

b. Is r statistically significant?   Yes   No

c. Did you think r would be higher? Why or why not?

9. In the preliminary section we also saw that the more modern nations were less racially prejudiced than the less modern nations. Are they also less anti-Semitic?

> Data File: **NATIONS**
> Task: **Scatterplot**
> ➤ Dependent Variable: **58) ANTI-SEM.**
> ➤ Independent Variable: **143) % INDUS $**
> ➤ Display: **Reg. Line**

In the space below, summarize what the results of this scatterplot indicate.

10. In the United States, important population shifts occurred in the 1980s.

> ➤ *Data File:* **STATES**
> ➤ *Task:* **Mapping**
> ➤ *Variable 1:* **3) POP GROW**
> ➤ *View:* **Map**

What part of the country generally experienced the greatest population growth during the 1980s? _____

11. Some states gained in population, while others lost population.

> *Data File:* **STATES**
> *Task:* **Mapping**
> *Variable 1:* **3) POP GROW**
> ➤ *View:* **List: RANK**

a. Which state had the highest population growth? _____

b. Which state lost the highest percent of its population? _____

c. What ranking did your home state achieve? _____

12. In the preliminary section we saw that attitudes on sexual matters have generally become more tolerant in the United States in the last 25 years. What about views on sex education? Use the HISTORY data set and HISTORICAL TRENDS task to obtain a graph of changes during this period in the percent approving of sex education in the schools. (Hint: When selecting variables, use the [Search] feature to locate the appropriate item.)

a. Is approval for sex education now higher than in the early 1970s, lower, or about the same? (Circle one.)

Higher

Lower

About the same

Exercise 16: Social Change and Modernization

b.   From the late 1970s to the late 1980s, did approval for sex education rise,
     fall, or stay about the same? (Circle one.)   Rise

                                                                                    Fall

                                                                              Stay the same
c.   What might account for the changes depicted in the graph?

APPENDIX A: STUDENT ExplorIt REFERENCE SECTION

This appendix provides additional information on using Student ExplorIt. If you have not already done so, read the instructions in *Getting Started* at the beginning of this book.

The first part of Appendix A provides information on *Student ExplorIt for Windows 95*. Little information is provided here because most of it is available via the on-line help. The second part of Appendix A provides information on *Student ExplorIt for DOS*. This section is substantially longer because on-line help is not provided in the DOS version of the program.

STUDENT EXPLORIT FOR WINDOWS 95

INSTALLATION

The instructions for installing *Student ExplorIt for Windows 95* have been provided in the *Getting Started* section of this workbook. If you have problems installing Student ExplorIt, first check to see that your computer has the minimum system requirements. (Note: *Student ExplorIt for Windows 95* does not work on computers running Windows 3.0 or 3.1. You should instead use *Student ExplorIt for DOS*.)

During the installation, you are asked whether you want the data files (e.g., GSS, STATES, etc.) installed on your hard drive. If you choose *not* to install the data files on your hard drive, you must insert the 3.5" diskette (which contains the data files) every time you run Student ExplorIt. If you later change your mind about the data file option you want to use, you will need to reinstall the program.

ON-LINE HELP AND SOFTWARE QUESTIONS

Student ExplorIt for Windows 95 offers extensive on-line help. You can obtain task-specific help by pressing **[F1]** at any point in the program. For example, if you are performing a scatterplot analysis, you can press **[F1]** to see the help for the SCATTERPLOT task.

If you prefer to browse through a list of the available help topics, select **Help** from the pull-down menu at the top of the screen and select the **Help Topics** option. At this point, you will be provided with a list of topic areas. Each topic area is represented by a closed-book icon. To see what information is available in a given topic area, double-click on a book to "open" it. (For this version of the software, use only the "Student ExplorIt" section of help; do not use the "Student MicroCase" section.) When you double-click on a book icon, a list of help topics is shown. A help topic is represented by a piece of paper with a question mark on it. Double-click on a help topic to view it.

If you have questions about *Student ExplorIt for Windows 95*, the first step is to try the on-line help described above. If you are not very familiar with software or computers, a second option is to ask a classmate or your instructor for assistance. If you are still unable to get an answer to your question or to resolve a problem, go to the Technical Support section of MicroCase's web site at http://www.microcase.com/student/support.html.

STUDENT ExplorIt FOR DOS

If you have not already done so, read the instructions in *Getting Started* at the beginning of this book. This appendix provides additional information on using *Student ExplorIt for DOS*. (If you are using *Student ExplorIt for Windows 95*, refer to the previous page.)

GETTING YOUR MOUSE WORKING

If you are using *Student ExplorIt for DOS* on a Windows 3.0/3.1 computer and the mouse arrow fails to appear on your screen, follow these instructions for loading the mouse driver:

1. Exit Windows. (You can exit by using <Alt> <F4> to close each window.)

2. At the DOS prompt (C:\>), type **MOUSE** and press <Enter>.

3. If you get an error message (e.g., "Bad command or file name," "Invalid directory"), try typing either **C:\MOUSE\MOUSE** and pressing <Enter> or **C:\LMOUSE\MOUSE** and pressing <Enter>.

4. If you get a message that tells you the mouse driver has been loaded, type **A:EXPLORIT** and press <ENTER> to start the program. If the mouse works now, you will need to follow the same procedure for loading the mouse driver each time you want to run the ExplorIt program.

5. If you still cannot see the mouse on the screen, contact the manufacturer of your computer to obtain a DOS mouse driver.

6. To return to Windows, type **WIN** and press <ENTER>.

HELP AND SOFTWARE QUESTIONS

If you encounter problems with *Student ExplorIt for DOS*, the first step is to review relevant sections in this workbook that might address your problem, such as this appendix. Next, you should try the help instructions provided at the bottoms of most screens. If you are not very familiar with software or computers, a third option is to ask a classmate or your instructor for assistance. If you are still unable to get an answer to your question or to resolve a problem, go to the Technical Support section of MicroCase's web site at http://www.microcase.com/student/support.html.

STANDARD OPERATIONS

These operations appear on the top row of onscreen buttons in *Student ExplorIt for DOS* (not all buttons appear on all screens):

EXIT

When you are within a task, clicking the [Exit] button will return you to the variable selection screen for the task. If you are at the variable selection screen within a particular task, this button will return you to the main menu. If you are at the main menu, you can exit the program by clicking the [Exit] button.

PRINT

To print the information shown on the screen, click the [Print] button. A window will open giving the current settings for the printer. If the settings are correct, click [OK] to send the results to the printer. If you change your mind about printing something, click [Cancel]. If the printer settings are incorrect, click [Settings] and follow the instructions in the next paragraph. If you want to save results to a disk file and print it later, use the [Disk File] option. In this case, type in the directory and file name you want to use (using standard DOS conventions) and click [OK]. This option will save text screens, but *not* graphics screens, to the named file. Also use the [Disk File] option if your printer is not supported by *Student ExplorIt for DOS*. Such files can be opened and printed using any word processor program.

Changing Printer Settings

To change the printer settings, click [Print] and then choose [Settings]. To select a new printer from the list, click its name and then click [OK]. If your printer doesn't appear on the list, try each printer and use the one that works best. The [Text (ASCII)] option will not allow you to print graphics.

You are also given the option of directing the printer to a particular port (the physical connector from your computer to your printer). Use LPT1: unless you have been instructed otherwise. (If you are printing over a network and are unable to print, check with your instructor or network administrator for additional information.)

VARIABLES

The [Variables] button will let you look at the list of variables for the data file that is currently open. The description of the highlighted variable is shown in a box at the lower right of the window. You can move this highlight through the list of variables by clicking on the scroll buttons to the right of the list box, by using the up and down cursor keys (as well as the <PgUp>, <PgDn>, <Home>, and <End> keys), or by clicking a variable once with the mouse.

You can also search the variable names and descriptions for a word, a partial word, or a phrase. Perhaps, for example, you want to find a variable about income. Click the [Search] button. Type **income** and press <Enter> (or click [OK]). The variable list window now contains only the variables that have the word *income* in either the variable name or the variable description. To return to the full list of variables, click [Full List].

SELECTING A SUBSET

Most statistical analysis tasks in Student ExplorIt allow you to analyze a subset of cases. For example, if you wanted to look only at females in the GSS, you would need to select a subset.

To limit your analysis to a subset of cases, select the primary variables for analysis as usual. But before you click [OK] to see the final results, click the [Subset Variables] button. A subset selection screen will appear and you may select up to four subset variables. You should select a subset variable as you would a regular variable. But upon choosing a variable, you will immediately be prompted to provide information on the *categories* to be used in selecting cases for the subset. Since there are two different types of variables, there are two different ways to select subset categories.

Selecting Subsets with Categorical Variables

A variable is a "categorical variable" when each of its categories has a discrete name or label (e.g., "Male" or "Female" for the variable GENDER). After you select a subset variable, the screen will ask you to select a category for a subset. Click it once to move the highlight to it and then click a second time to select the category. An "x" will appear in the box to the left of the name. You may select as many or as few categories as you wish.

After you have selected the categories, you must indicate whether cases in the indicated categories are to be *included* or *excluded* in the analysis. For example, if you select GENDER as the subset variable and "Male" as the category, you may choose to "include" only males in the analysis, or you may choose to "exclude" them and look only at females ("non-male"). Just click the option you want. Then click [OK] to return to the variable selection screen.

Selecting Subsets with Noncategorical Variables

If a variable uses a range of numbers for its values (e.g., 1.2–13.3) and these values do not fall into discrete categories (such as "male" or "female"), a different method of selecting cases is used. After you select the subset variable, the screen will ask for the low and the high value to be used in selecting the cases. For example, if you type 1.2 as the low value and 6.7 as the high value, cases with values between 1.2 and 6.7 on the subset variable will be included in the analysis. Then click [OK] to return to the variable selection screen.

MULTIPLE SUBSET VARIABLES

If you want to base the subset on more than one variable, select a second variable and define another subset. For example, if you wanted to include *only* white females, you would define your subset using two subset variables: WHTE/AFRAM (include category white) and GENDER (include category female).

DELETING SUBSETS

All variable selections, including selections of subset variables, are saved by Student ExplorIt after you exit from a results screen. If you are going to conduct another analysis using the same task, it is important to delete or clear subset variables when you are finished with them. There are three ways to accomplish this. First, you can click the [Clear All] button on the variable selection screen, which will clear all selected variables. Second, you can click the [Subset Variables] button and then click [Delete] to eliminate each subset variable individually. Or, third, you can return to the main menu, which will automatically deletes all variable selections.

AUTO-ANALYZER

SELECTING VARIABLES

Select an analyzer variable according to the variable selection instructions in *Getting Started*, at the beginning of this book. Click [OK] to continue.

VIEWS

Univariate: Auto-Analyzer first shows the overall distribution of the selected variable. In addition, a brief description of this distribution is shown below the table.

Demographic Views: Nine demographic views are available from Auto-Analyzer: Sex, Race, Political Party, Marital, Religion, Region, Age, Education, and Income. Each option shows the table linking the selected variable and the demographic variable, with proper percentages. In addition, a textual description of the table is provided. If the description contains a scroll bar, use the mouse or cursor keys to see the additional information.

Summary: This option includes the textual summaries for each of the nine demographic variables and the distribution for the primary variable.

UNIVARIATE STATISTICS

SELECTING VARIABLES

Select a primary variable according to the variable selection instructions in *Getting Started* at the beginning of this book. If you want a subset of cases, see the *Selecting Subsets* section toward the beginning of this appendix. (If you need to erase all existing variable information for this task, click the [Clear All] button.)

Once your variables have been selected, click [OK] to obtain the results of your analysis.

VIEWS

These options are located on the top row of buttons on the results screen:

Pie: This shows the distribution of the variable and displays it in the form of a pie chart.

Bar [Frequency]: This displays the distribution of cases in the form of a bar graph. Information on the first category of the variable is shown below the bar graph. To see information on other categories, click the bar of your choice.

Bar [Cumulative]: This displays the cumulative percentage of cases for the category displayed.

Statistics: This button produces the frequency, percent, cumulative percent, and z-score for each category of the variable. Summary statistics are shown at the top of the screen.

CROSS-TABULATION

SELECTING VARIABLES

Select a row variable and a column variable according to the variable selection instructions in *Getting Started* at the beginning of this book. If you want a subset of cases, see the instructions on selecting subsets at the beginning of this appendix. (If you wish to erase all existing variable information for this task, click the [Clear All] option.)

Control Variables (optional)

The [Control Variables] option can be selected from the variable selection screen in the CROSS-TABU-LATION task. A window for selecting control variables will open. At this point you may select up to three control variables using the usual method of selecting variables. When you have selected the control variables for your analysis, click [OK] to return to the variable selection. When the final results are displayed, a "partial table" is shown for each level of the control variable(s).

Once your variables have been selected, click [OK] to obtain the results of your analysis.

VIEWS

Table: This shows the cross-tabulation results for the selected row and column variables. If one or more control variables have been selected, the categories of the cases contained in this subtable are shown at the top of the screen. If the entire table will not fit on the screen, use the cursor keys or click the scroll bars at the bottom and/or right of the table to scroll through additional rows and columns.

Display (these options are located in the menu at the right):

Column %: This option provides the column percentages for the table. Missing data are ignored.

Row %: This option provides the row percentages for the table. Missing data are ignored.

Total %: This option provides the total percentages for the table. Missing data are ignored.

Freq.: This option returns the table to the original frequencies.

Previous (Optional): If a control variable has been selected, this option will let you return to previously viewed subtables.

Next (Optional): If a control variable has been selected, this option will take you to the next subtable.

Stats: This view shows the summary statistics for the table.

Bar: This shows a stacked bar chart representing the table.

Collapse: The [Collapse] button appears whenever a cross-tabulation table is showing. This option allows you to combine categories of a table or to drop categories entirely. (No changes using the [Collapse] option permanently modify the variable—the changes disappear as soon as you leave the results screen.) To select the categories to be combined or dropped, click the category labels (this causes the entire row or column to be highlighted). Then click the [Collapse] button. Here you are given the choice of creating a new collapsed category or turning the highlighted categories into missing data. To create a new collapsed category, type in a category name and click [OK]. If you want to drop the highlighted categories, click the [Drop] button. The table will then be updated according to your selections. The [Collapse] option will work with both the row and column variables in a cross-tabulation table, although only one variable at a time can be modified.

MAPPING

SELECTING VARIABLES

Select a variable according to the variable selection instructions in *Getting Started*, at the beginning of this book. If you want to use a subset of cases, see the instructions on selecting subsets at the beginning of this appendix. If you wish to erase all existing variable information for this task, click the [Clear All] option.

Once you select your variables, click [OK] to obtain the map.

VIEWS

These options are located on the top row of buttons.

Map: The selected variable is mapped into five different levels from light (lowest) to dark (highest). Cases for which data are unavailable are left blank.

<u>**Display**</u> (these options are located at the right side of the top button row):

Legend: This option provides the values represented by each level, known as the map legend, in a box at the lower right. Selecting this option a second time removes the legend.

Find: This option allows you to locate a particular case. When you select this option, a list of all cases will be shown. Click twice on the case you want to select so that an "x" appears to the left of the name. If the case you want to select is not shown in the window, scroll to it using the cursor keys or click the scroll bar at the right side of the window. Click [OK] to continue. The selected case is highlighted on the map and its name, value, and rank are shown at the bottom of the screen. To deselect this highlighted case, click the [Find] button again or click outside the border of the map.

<u>Special Feature:</u> Click a case. You may use the mouse to click a case. The selected case is shown in the highlight color and its name, value, and rank are shown at the bottom of the screen. To deselect this case, click outside the border of the map.

Spot: This view shows each case as a filled circle. The size of the spot, or circle, is relative to the value of the case—cases with higher values have larger spots. The color of the spot is the same as that used in the Map view. To return to the original map, click the [Spot] button again.

List [Rank]: This option allows you to rank the cases on the variable from high to low. The name of the variable, its rank, and its value are displayed. In addition, the map color for the case is shown to the left of the name. Since the initial screen shows only the highest cases, use the scroll bar at the right side of the window to see additional values.

List [Alpha]: This option works the same as [List: Rank], except that cases are shown in alphabetical order.

SCATTERPLOT

SELECTING VARIABLES

Select two variables according to the variable selection instructions provided in *Getting Started*, at the beginning of this book. The dependent variable will be represented by the *y*-axis on the graph and the independent variable will be represented by the *x*-axis.

If you want a subset of cases, see the instructions on selecting subsets at the beginning of this appendix. (If you need to erase all existing variable information for this task, click the [Clear All] option.)

Once your variables have been selected, click [OK] to obtain the scatterplot results.

VIEWS

The scatterplot task has only one view.

The scatterplot graph shows each case as a dot, with the *x*-axis representing the independent variable and the *y*-axis representing the dependent variable. The correlation coefficient (r) is shown at the lower left of the screen. One asterisk indicates a .05 level of statistical significance; two asterisks, a .01 level.

Display (these options are located in the Subview menu at the left of the screen):

Reg.Line: This option places the regression line on the scatterplot graph and gives the equation for the regression line below the graph. Selecting this option a second time removes the line.

Residuals: The residual for each case—a vertical line from the case to the regression line—is shown on the scatterplot graph. Selecting this option a second time removes the residual lines.

Find Case: This option allows you to identify—place a box around—the dot representing a particular case. When you select this option, a list of all cases will be shown. If the case you want to see is not immediately visible in the window, scroll to the appropriate case using the cursor keys or click the scroll bar at the right side of the window. To select a case, click it twice so that an "x" appears in the box to the left. Click [OK] to return to the scatterplot graph. Information on the selected case is shown in the window at the lower left—this window is described in the Special Features section, below.

Outlier: This option will identify—place a box around—the dot representing the outlier. In this application, the outlier is defined as the case that would make the greatest change in the correlation coefficient if it were removed. When you select this option, information on the outlier case is given in a window at the lower left—this window is described in Special Features, below.

Special Features

Click a dot: You may use the mouse to click a dot. A box is placed around this dot and information on the case is shown in the window at the lower left.

X button: If you click the X button at the bottom of the graph, the description of the independent variable will be shown below the graph. Click the "x" in the upper right corner of the description to close this window.

Y button: If you click the Y button at the left side of the graph, the description of the dependent variable will be shown below the graph. Click the "x" in the upper right corner of the description to close this window.

Window: When certain displays are selected from the scatterplot screen, a window appears that contains information on the case currently highlighted in the scatterplot. This window will appear when you use the [Outlier] option or the [Find case] option, or if you click a dot in the scatterplot. The name of the case and its value on the independent (or x) variable and on the dependent (or y) variable are shown. In addition, you are given the option of removing the case from the scatterplot. If you click the option [Remove Case from Graph], the case will be removed from the scatterplot and any statistics will be recalculated with that case removed. (The removal of this case is only temporary—the original data set is not modified.) The bottom line of the window tells you what the value and statistical significance of the correlation coefficient will be if the highlighted case is removed. Click the "x" in the upper right corner of this window to close it.

APPENDIX B: VARIABLE NAMES AND SOURCES

Note for MicroCase Users: These data files may be used with MicroCase. If you are moving variables from these files into other MicroCase files, or vice versa, you may need to reorder the cases. Also note that files that have been modified in MicroCase will not function properly in Student ExplorIt.

◆ DATA FILE: GSS ◆

1) WORKING?
2) OCCUPATION
3) PRESTIGE
4) DAD OCC.
5) MA OCC.
6) EDUCATION
7) FAM INCOME
8) OWN INCOME
9) WELFARE $2
10) LIKE JOB?
11) WRK IF $$
12) CLASS?
13) SAT.$?
14) EVER UNEMP
15) UNIONIZED?
16) FEM JOB+
17) JOBS ALL
18) INEQUAL 3
19) PARS.PRESG
20) PARSDEGREE
21) HOMEMAKER?
22) INC. DIF?
23) AID UNEMP.
24) LABOR?
25) BLACK $
26) INTERMAR.?
27) RACE SEG.
28) BL.IN AREA
29) RAC PRES
30) BLKS IMP
31) RAC DIF2
32) RACE DIF4
33) WHITE WORK
34) BLACK WORK
35) WHITE IQ
36) BLACK IQ
37) MARRY BLK
38) IMM. CRIME
39) IMM. ECON

40) IMM. JOBS
41) LET IN1
42) WHTE/AFRAM
43) RACE/ETHNC
44) RACEGENDER
45) GENDER
46) WOMEN HOME
47) MOTH.WORK?
48) PRESCH.WRK
49) WIFE@HOME
50) SEX PROMO
51) CARE CHLD5
52) HIRE WOMEN
53) YOURSELF
54) MARITAL
55) # SIBS
56) # CHILDREN
57) AGE
58) AGE KD BRN
59) HOME AT 16
60) MA WRK GRW
61) ALONE?
62) REGION
63) HAP.MARR.?
64) SOC.KIN
65) SOC.NEIGH.
66) SOC.FRIEND
67) LIVE W KID
68) ABORT ANY
69) IDEAL#KIDS
70) SEX ED?
71) DIV.EASY?
72) PREM.SEX
73) XMAR.SEX
74) HOMO.SEX
75) PORN.LAW?
76) SPANKING
77) EUTHANASIA
78) LIVE WITH

79) HAPPY REL
80) WILL WED1
81) TRAD/MOD
82) FAM LIFE
83) BAL WK/FAM
84) CLSENEI
85) CLSETOWN
86) SX.PRTNRS?
87) SEX FREQ.
88) CONDOM
89) SOUTH
90) AGE 65+
91) DIVORCED
92) PARS. DIV?
93) THINKSELF
94) URBAN?
95) KID LEARN
96) THINK/OBEY
97) ANGRY FAM
98) MISS HOLS?
99) POL.PARTY
100) VOTE IN 92
101) WHO IN 92?
102) POL. VIEW
103) ATHEIST SP
104) RACIST SP
105) COMMUN.SP
106) HOMO. SP
107) FED.GOV'T?
108) SUP.COURT?
109) CONGRESS?
110) POL.INTR.
111) PROUD DEM
112) GOV.POW.
113) R PROTEST1
114) R PROTEST2
115) R PROTEST3
116) R PROTEST4
117) EXECUTE?

118) GUN LAW?
119) GRASS?
120) SUIC.ILL
121) FEAR WALK
122) OWN GUN?
123) SELL SEX
124) OBEY LAW
125) SP.POLICE
126) RELIGION
127) ATTEND
128) PRAY
129) RELIG. 1ST
130) BIBLE
131) DENOM
132) HAPPY?
133) HEALTH
134) LIFE
135) HELPFUL?

136) ADVANTAGE?
137) TRUSTED
138) MEDICINE?
139) GOV.MED.
140) SHAKE BLUE
141) LONELY
142) NO PLAN
143) EVER MENTL
144) POL.EFF.11
145) GOOD LIFE
146) #CRCT.WORD
147) GENE GOOD
148) SELL ORGAN
149) GET AHEAD?
150) NEWSPAPER?
151) WATCH TV
152) VOL HEALTH
153) VOL EDUC

154) VOL HUMAN
155) VOL ENVIR
156) VOL YOUTH
157) GO DOC
158) EAT OUT
159) VOL RELIG
160) SEE FILM
161) I-SEX
162) I-RACE
163) I-POL PRTY
164) I-MARITAL
165) I-RELIGION
166) I-REGION
167) I-AGE
168) I-SCHOOL
169) I-INCOME

◆ DATA FILE: NATIONS ◆

1) COUNTRY
2) POPULATION
3) DENSITY
4) URBAN %
5) URBAN GRWT
6) POP GROWTH
7) NETMIGRT
8) BIRTH RATE
9) FERTILITY
10) LARGE FAML
11) INF. MORTL
12) MOM MORTAL
13) CONTRACEPT
14) ABORTION
15) ABORT LEGL
16) MOM HEALTH
17) AB. UNWANT
18) DEATH RATE
19) LIFE EXPCT
20) SEX RATIO
21) ECON DEVEL
22) QUAL. LIFE
23) CALORIES
24) MEAT CONS.
25) %STUNTED

26) GREENHOUSE
27) NAT. PROD.
28) ECON GROW
29) $ PER CAP
30) UNEMPLYRT
31) ELECTRIC
32) % AGRIC $
33) % IN AGR.
34) CARS/1000
35) TV 1000
36) RADIO 100
37) PHONE 1000
38) MOVIES
39) NEWSPAPER
40) EDUCATION
41) % GO 5TH
42) %MUSLIM
43) %CHRISTIAN
44) %CATHOLIC
45) %HINDU
46) %BUDDHIST
47) %JEWISH
48) JEHOV.WITN
49) MORMONS
50) REL.PERSON

51) CH.ATTEND
52) GOD IMPORT
53) PRAY?
54) REINCARNAT
55) MULTI-CULT
56) C.CONFLICT
57) DEMOCRACY
58) ANTI-SEM.
59) ANTI-FORGN
60) ANTI-MUSLM
61) RACISM
62) ANTI-GAY
63) HOME LIFE?
64) HAPPY SEX?
65) CHORES?
66) FRIENDS?
67) LEISURE?
68) FEM.PROF.
69) FEM.MANAGE
70) FEM.OFFICE
71) %FEM.LEGIS
72) %FEM.HEADS
73) F/M EMPLOY
74) M/F EDUC.
75) GENDER EQ

76) FEM POWER
77) SEX MUTIL
78) SINGLE MOM
79) WORK MOM?
80) HOME&KIDS
81) WED PASSE'
82) REVOLUTION
83) % LEFTISTS
84) %RIGHTISTS
85) LEFT/RIGHT
86) P.INTEREST
87) PETITION?
88) BOYCOTT?
89) DEMONSTRAT
90) TALK POL.
91) WORK PRIDE
92) WORK IMPT?
93) UNIONIZED?
94) UNIONS?
95) POOR LAZY
96) INJUSTICE
97) WORKER OWN
98) TRUST?

99) TRUST KIN?
100) TRUST CITZ
101) CHEAT GOVT
102) CHEAT BUS$
103) CHEAT TAX
104) TAKE BRIBE
105) LYING
106) HOT BUY
107) LITTERING
108) COP CONFID
109) EX-MARITAL
110) MINOR SEX
111) GAY SEX
112) PROSTITUTE
113) CIRRHOSIS
114) SUICIDE
115) SUICIDE NO
116) EUTHANASIA
117) AIDS
118) DRUGS
119) SMOKE DOPE
120) ALCOHOL
121) SPIRITS

122) BEER DRINK
123) WINE DRINK
124) CIGARETTES
125) DO SPORTS?
126) VERY HAPPY
127) NATL PRIDE
128) FAMILY IMP
129) KID MANNER
130) KID INDEPN
131) KID OBEY
132) KID THRIFT
133) REGION
134) SUICIDE OK
135) MALE LIFEX
136) FEM LIFEX
137) DEMOCRACY2
138) PETITION2
139) POL.INTRST
140) TALK POLT
141) BOYCOTT2
142) DEMONSTRA2
143) % INDUS $
144) % PROT

♦ DATA FILE: STATES ♦

1) FED LAND
2) WARM WINTR
3) POP GROW
4) POPULATION
5) $>AGE 64
6) POP<AGE18
7) AVG. AGE
8) DENSITY
9) %MALE
10) %FEMALE
11) %URBAN
12) %RURAL
13) %WHITE
14) %AFRIC.AM
15) %AMER.IN
16) %ASIAN/P
17) %IRISH
18) %ITALIAN
19) %POLISH
20) %SCOTTSH
21) %CHINESE

22) %JAPANESE
23) %HISPANIC
24) %MEXICAN
25) %PUERTO.RC
26) %CUBAN
27) IMMIGRANTS
28) DIFF.STATE
29) %NO RELIG.
30) %JEWISH
31) %CATHOLIC
32) %BAPTIST
33) CH.MEMBERS
34) SUICIDE
35) MARRIAGES
36) DIVORCES
37) %SINGLE
38) %DIVORCED
39) HH CHANGE
40) F HEAD W/C
41) BIRTHS
42) MA <AGE15

43) MA <AGE20
44) CHLD MORTL
45) HEART DTHS
46) ONE+/ROOM
47) NEW HOMES
48) NO PHONES
49) % NO CARS
50) %IN PRISON
51) %NURS.HOME
52) %MENTAL.HS
53) %EMERG.SHL
54) %IN STREET
55) %POOR
56) %POOR>65
57) %WHT.POOR
58) %BLK.POOR
59) %AM.IND PR
60) %ASIAN PR
61) %HISP.PR
62) %POOR FAM.
63) %CHLD POOR

64) %FEM.HEAD	82) COSMO	100) ASSAULT
65) MED.FAM. $	83) COKE USERS	101) BURGL
66) PER CAP. $	84) ALCOHOL	102) LARCENY
67) UNEMPLOYED	85) BEER	103) CAR THEFT
68) PUB.TRNS	86) WINE	104) KID ABUSE
69) AVG. TRAVL	87) %WINE	105) PICKUPS
70) %AGRI.EMP	88) %BEER	106) %CLINTON92
71) %MINE EMP	89) %FAT	107) %BUSH 92
72) %MANUF.EMP	90) SYPHILIS	108) %PEROT 92
73) SOME H.S.	91) AIDS	109) %TV>6HRS
74) HIGH SCH.	92) HEALTH.INS	110) JR LEAGUE
75) COLL.DEGR.	93) %FEM MD	111) BOYS LIFE
76) DROPOUTS	94) MDS	112) SURVIVE
77) ABORTION	95) V.CRIME	113) MOBILITY
78) S.ACCENT	96) P.CRIME	114) NOT DENSE
79) PLAYBOY	97) MURDER	115) %FEMALE LG
80) F&STREAM	98) RAPE	116) HMWK>2
81) GOURMET	99) ROBBERY	117) STATE NAME

SOURCES

GSS

The GSS data file is based on selected variables from the National Opinion Research Center (University of Chicago) General Social Survey for 1996, distributed by the Roper Center and the Inter-University Consortium for Political and Social Research. The principal investigators are James A. Davis and Tom W. Smith.

NATIONS

The data in the NATIONS file are from a variety of sources. The variable description for each variable uses the following abbreviations to indicate the source.

CA: Church Almanac, 1995–1996, Salt Lake City Deseret News

FITW: Freedom in the World, 1995, Freedom House

HDR: Human Develoment Report, 1993 and 1995, United Nations Development Program

IP: International Profile: Alcohol and Other Drugs, 1994, Alcoholism and Drug Addiction Research Foundation

NBWR: The New Book of World Rankings, 3d Edition, 1991, Facts on File

PON: The Progress of Nations, 1996, UNICEF

SAUS: Statistical Abstract of the United States, 1992, U.S. Department of Commerce

STARK: Coded and calculated by Rodney Stark

TWF: The World Factbook, 1994, Central Intelligence Agency

TWW: The World's Women, 1995, United Nations

WABF: The World Atlas and Book of Facts, 1995, World Almanac Books

WCE: World Christian Encyclopedia, David B. Barrett, editor, Oxford University Press, 1982

WVS: World Values Survey, 1981–1984, 1990–1993, Institute for Social Research, Inter-university Consortium for Political and Social Research

YJW: The Yearbook of Jehovah's Witnesses, 1995

STATES

The data in the STATES file are from a variety of sources. The variable description for each variable uses the following abbreviations to indicate the source.

ABC: Blue Book, Audit Bureau of Circulation

CENSUS: The summary volumes of the 1990 U.S. Census

CHURCH: Churches and Church Membership in the United States, Glenmary Research Center

DES: Digest of Education Statistics, U.S. Dept. of Education

HCSR: Health Care State Rankings, Morgan Quitno

HEALTH: Health, Centers for Disease Control

HIGHWAY: Highway Statistics, Federal Highway Administration, U.S. Dept. of Transportation

KOSMIN: Kosmin, Barry A. 1991. Research Report: The National Survey of Religious Identification. New York: CUNY Graduate Center.

MMWR: Morbidity and Mortality Weekly Report, Centers for Disease Control

MVSR: Monthly Vital Statistics Report, Centers for Disease Control

NCHS: National Center for Health Statistics

S.A.: Statistical Abstract of the United States

SMAD: State and Metropolitan Area Data Book, 1991, U.S. Dept. of Commerce

S.P.R.: State Policy Reference

UCR: The Uniform Crime Reports, U.S. Dept. of Justice

WA: World Almanac

LICENSE AGREEMENT

READ THIS LICENSE AGREEMENT CAREFULLY BEFORE OPENING THE SEALED SOFTWARE PACKAGE. BY OPENING THIS PACKAGE YOU ACCEPT THE TERMS OF THIS AGREEMENT.

MicroCase® Corporation, hereinafter called the Licensor, grants the purchaser of this software, hereinafter called the Licensee, the right to use and reproduce the following software: **Discovering Sociology: An Introduction Using ExplorIt,** in accordance with the following terms and conditions.

Permitted Uses

◆ You may use this software only for educational purposes.

◆ You may use the software on any compatible computer, provided the software is used on only one computer and by one user at a time.

◆ You may make a backup copy of the diskette(s).

Prohibited Uses

◆ You may not use this software for any purposes other than educational purposes.

◆ You may not make copies of the documentation or program disk, except backup copies as described above.

◆ You may not distribute, rent, sub-license, or lease the software or documentation.

◆ You may not alter, modify, or adapt the software or documentation, including, but not limited to, translating, decompiling, disassembling, or creating derivative works.

◆ You may not use the software on a network, file server, or virtual disk.

THIS AGREEMENT IS EFFECTIVE UNTIL TERMINATED. IT WILL TERMINATE IF LICENSEE FAILS TO COMPLY WITH ANY TERM OR CONDITION OF THIS AGREEMENT. LICENSEE MAY TERMINATE IT AT ANY OTHER TIME BY DESTROYING THE SOFTWARE TOGETHER WITH ALL COPIES. IF THIS AGREEMENT IS TERMINATED BY LICENSOR, LICENSEE AGREES TO EITHER DESTROY OR RETURN THE ORIGINAL AND ALL EXISTING COPIES OF THE SOFTWARE TO THE LICENSOR WITHIN FIVE (5) DAYS AFTER RECEIVING NOTICE OF TERMINATION FROM THE LICENSOR.

MicroCase Corporation retains all rights not expressly granted in this License Agreement. Nothing in the License Agreement constitutes a waiver of MicroCase Corporation's rights under the U.S. copyright laws or any other federal or state law.

Should you have any questions concerning this Agreement, you may contact MicroCase Corporation by writing to MicroCase Corporation, 14110 N.E. 21st Street, Bellevue, WA 98007, Attn: College Publishing Division.